T0259419

Modeling and Optimization of Fermentation Processes

Process Simulation and Modeling, 1

Modeling and Optimization of Fermentation Processes

B. VOLESKY

McGill University, Montreal, Quebec, Canada

and

J. VOTRUBA

Czechoslovak Academy of Sciences, Prague, Czechoslovakia

ELSEVIER
Amsterdam — London — New York — Tokyo   1992

ELSEVIER SCIENCE PUBLISHERS B.V.
Sara Burgerhartstraat 25
P.O. Box 211, 1000 AE Amsterdam, The Netherlands

ISBN: 0-444-89588-4

This book is printed on acid-free paper.

Printed and bound by Antony Rowe Ltd, Eastbourne
Transferred to digital print on demand, 2005

CONTENTS

PREFACE

The ancient alchemists stopped hammering at the iron rod in attempts to convert it into gold when solid scientific predictions taught them the impossibility of the task. Much too often, however, it seems that we are behaving in a similar way in our attempts to make the cell synthesize valuable products for us. To run a fermentation process is still more of an art than an exact science. We should not be fooled by the contemporary advances of "biotechnology"; it is based on almost as empirical and experimental approach as that practiced by the good old alchemist some three or four centuries ago. It is the power of scientifically based predictions that leads us to the best results and the optimum process configurations.

It is the true industrial bio-TECHNOLOGY that this book attempts to address, bringing into it some basic rudimentary methods of process description and optimization based on the magic of the mathematical equation. This does not mean that the bio-scientists among us should stop reading at this point. Simple differential calculus is the backbone of this volume. The concept of mass balancing is summarized in Part II for those who are not used to this most useful and classical basic engineering tool. An extensive and descriptive case study of a selected bio-process in Part III elucidates the concepts of very pragmatic mathematical modeling of the bioreactor systems outlined in Part I.

While the individual concepts dealt with in this volume are of a rather basic nature to the specialists in individual areas involved, it is actually the interdisciplinary nature of the bioprocess field that presents the challenge. Recent advances in these individual areas now make it possible to approach the exciting interdisciplinary task with reasonable confidence. The accumulated knowledge of biochemical microbial pathways, and the experience with description and optimization of chemical reactors, developed in the last three or four decades, is catalyzed by the contemporary power of small, extremely fast and accessible computers loaded with software of powerful mathematical routines. The result is a scientific environment where a qualitative leap can be taken in attempts to quantify some bio-catalytic processes; the industrial ones being of a special interest.

This volume is meant for those who are dealing with the bio-process elements in the laboratory or on a large scale. It is meant for the engineer as well as for the science student, because it is in between the classical fields where the interdisciplinary challenge is, and where the opportunity beckons. Forget the traditional boundaries of scientific disciplines you were inoculated with at school and enjoy the new world of interdisciplinary excitement. It is the energies of this excitement that will bring you through this volume and into the new world of more exact scientific and technological endeavours in biotechnology. It is perhaps time to prepare to leave the age of (bio)-alchemy. The old saying is that every long trek starts with the first step. We hope that this book could be your first step.

The Authors.

PART I

MATHEMATICAL MODELING
OF
MICROBIAL PROCESSES

1.   INTRODUCTION

For a proper design and operation of fermentation processes, microbial ore leaching, biological wastewater treatment, bioconversion of solar energy by (green) microorganisms, and for other numerous and varied processes of biotechnology, it is essential to know and to be able to quantitatively describe the key process variables relevant to the system kinetics. Such information serves as a basis for deriving an optimal process design and for developing its optimal operation. While in the chemical reactors the process kinetics reflects the reaction rates on a molecular level, microbial process dynamics is a result of relationships between the living microbial cell and its environment affecting the biochemical-physiological activity of the microbial population and thus the results of the whole bio-process. Dynamics and efficiency of the microbial process can be manipulated by the choice of microbial culture and by the physico-chemical environmental factors. An optimal bio-process results from combining the best choices in both areas. Without a suitable microbial strain it is not possible to realize the desired process and, similarly, by using inappropriate process conditions only very low product yields can be obtained even when high-production strains are employed. The methodology of strain selection, its genetic manipulation and optimization of production parameters has been traditionally based on very extensive experimental work, diametrically different from the methods of engineering optimization of a process with regard to its operating parameters.

Considering extremely high costs of industrial-scale experimentation, microbial process engineering approach to experimental work can make an efficient use of laboratory-scale experimentation employing a scaled-down "model system". A geometrically scaled-down copy of the process equipment or a sequence of operations can serve as a model to study the process. This study can be facilitated by developing an analogy of the system, for instance an electrical analogy, or even a mathematical abstraction (mathematical model) enabling simulation of the behaviour of an actual process by computations. In most cases, every model represents a certain approximation of the real system and represents a compromise between the high costs and complexity of experimentation with a real large-scale system and the ease of carrying out a smaller-scale experimental study.

This text will mainly deal with problems associated with the application of mathematical modeling methods as a tool of systems analysis in the field of biotechnology. The text is divided into major sections dealing respectively with the methodology of composing mathematical models of bioreactor performance, the types of material balances pertaining to the bioreactor system and, eventually, elaborating on the principles discussed, there is a comprehensive case study where different bioreactor arrangements are modelled and computer simulation of their performance demonstrated. Numerous examples and problems solved throughout the text make the comprehension of the concepts dealt with easier to understand and absorb.

The text has been prepared with broad and interdisciplinary readership in mind. The process engineer will find the concepts more familiar, however, his biochemical and microbiological background has to have been sufficiently developed. The biologist, on the other hand, needs to have basic preparation in integral and differential calculus and the section on mass balances will smooth his entry into the basic area of mathematical modeling of biosystems. In either case, open minded interdisciplinary curiosity, pragmatic approach and unsuppressible desire to be at the "cutting edge" of the contemporary development in new and rapidly expanding areas of biotechnology are the basic requisites for enjoying this text which is to assist in further development and application of the powerful methodology for study and optimization of bioreactor systems in the laboratory as well as in large-scale operation. After all, it is the technology component which is to ultimately fulfill the promises and expectations of the fascinating, new and highly interdisciplinary field of biotechnology.

2. SYSTEMS ANALYSIS APPROACH TO THE MATHEMATICAL MODELING OF FERMENTATION PROCESSES

Systems analysis is a basic method for description of complex phenomena and interactions among observed variables in the process under study[17]. The unified strategy for analysis of an arbitrary process determines the strategy for process optimization. By the process systems analysis we refer to the application of scientific methods to the recognition and definition of process-related problems and the development of procedures for their solution. In practice, for a fermentation process system, this approach is represented by several basic steps:

a) mathematical specification of the problem for the given physicochemical, biochemical and physiological conditions;

b) detailed strategy development resulting in obtaining adequate mathematical model(s) representing the given process;

c) synthesis of results and design of the optimization strategy for process control.

The biological process denotes an actual series of operations and interactions of non-living materials with living matter. Figure 2.1 presents a simplified summary of interactions and links between "microbial process engineering" and other science branches.

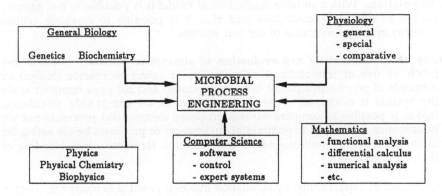

Figure 2.1: The place of "Microbial Process Engineering" among the established science disciplines

Process engineering mostly deals with observed macroscopic kinetics and stoichiometry of biological processes. The kinetics and stoichiometry are based on physiological studies but the theoretical background is developed from enzyme kinetics, metabolic pathways and sometimes on the basis of genetic laws.

The process system state variable is the quantity which can represent an imaginary coordinate in the "state space". Such variables can be determined either by direct measurement (biomass concentration, temperature, pH, etc.) or it can be of an indirect nature. That means that its value is calculated from other measured variables (yield coefficient, RQ, specific growth rate, etc.).

The state of the system is determined by the set of system variables and their corresponding respective rates of changes. Process parameter is a property of the system or its environment that can be assigned arbitrary numerical or linguistic values; also it is a constant or a coefficient in an equation often based on and derived from some assumption such as "ideally mixed", "normal behaviour", "loss of viability", etc.

Simulation is the study of the system or its part(s) by manipulation of its mathematical representation or its usually smaller physical model.

Process analysis involves an examination of the overall process, alternative technological variants and also eventually their economics. There are two main tests in the biotechnological industry with which engineers are ultimately concerned: the optimal operation of an existing plant and the design of new or modified technologies. In the area of operations, both control and optimization of the system performance stand out as two of the main functions of great concern to the process engineer. From a general viewpoint, systems analysis and process simulation have the following benefits:

A) **Extrapolation.** With a suitable mathematical model it is possible to test extreme ranges of operating conditions and also it is possible to establish critical patterns in the performance of the real process.

B) **Study of commutability and evaluation of alternative policies.** New factors (such as use of immobilized cell reactor, or novel bioreactor design) or elements of process equipment can be introduced and old ones removed while the system is examined to see if these changes are compatible. Simulation makes it possible to compare various proposed designs and processes not yet in operation and to test hypotheses about systems or processes before acting (as in the case of the continuous-flow cell retention fermentor with bleeding of the whole broth).

C) **Replication of experiments** by simulation makes it possible to study the effect of changes of system variables and process parameters.

D) **Test of sensitivity and system stability** to disturbances in basic process parameters can be examined.

E) **Optimal control and economic experimentation** can be studied leading to the optimal process design quickly and economically. A study of this sort with a real plant would be extremely risky, expensive and cumbersome involving costly large scale experimentation and design changes.

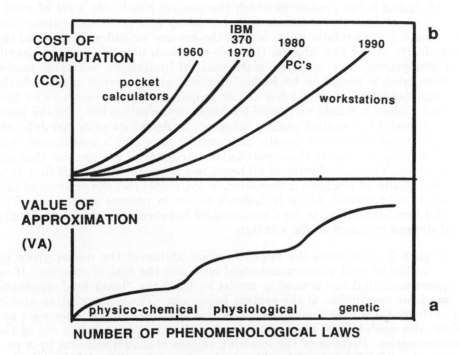

Figure 2.2: The value of approximation by the mathematical model (a) increases with its complexity. The costs of solution (computation) (b) have been gradually decreasing as the computer power is on the rise.

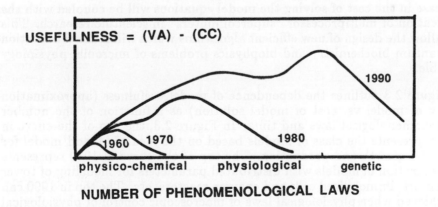

Figure 2.3: Model usefulness as a function of the number of phenomenological laws and time.

Modeling is the process in which the analyst constructs a set of mathematical relationships together with boundary and initial conditions that are isomorphic to relationships among the process variables. Because of the complexity of the real process (physico-chemical, physiological, biochemical and genetic laws) and the mathematical limitations, whatever model is developed is bound to be highly idealized and generally gives a faithful representation of only a few of the properties of the process. The first model is often a simple version of the mass conservation law. On the basis of this model the analyst usually attempts to detect its principal deficiencies. Several models are usually composed before one is established that satisfactorily represents those particular attributes of the process that are of interest. The typical error of all beginners in model building is that they try to include all available information in the model and the product of this effort is a "monster" which is difficult to use in process simulation. The model formulation has to be a compromise between reasonable complexity and desired economy of the solution.

Figure 2.2 compares the approximation ability of the model given by the number of used phenomenological laws and the cost of solution. If no phenomenological law is used in model building the "black box" approach is used for description of the system behaviour. The extrapolation ability of such model is very low. When only the physico-chemical laws were applied, the model could represent an optimal compromise in the era of the mid-seventies. Because of the lowering of cost of model solution by a new generation of software and hardware, the new generation of "physiological oriented" models represent the top of optimal model designs in the eighties. It is clear that further inclusion of genetic, biochemical and biophysical laws (eg. quantum biochemistry and molecular mechanics of protein action) will be in the center of interest of model designers in the upcoming era where the decrease in the cost of solving the model equations will be coupled with the application of multiprocessor "supercomputers" in scientific research. This will allow the design of new efficient algorithms which will make the solution of quantum biochemistry and biophysics problems of microbial physiology possible.

Figure 2.3 outlines the dependence of model usefulness (approximation ability of model vs. cost of model solution) as a function of the number of phenomenological laws and time. In Figure 2.3, the top of the curve in 1970 represents the class of models based on the idea of Monod model for microbial growth and production. The top of the curve in 1980 represents the generation of models with distributed parameters for modeling of tower fermentors, immobilized cells and enzyme reactors, etc. The top in 1990 can be achieved when physiological laws of macroscopic control of physiological functions will be applied. The top of usefulness at the end of the century means the implantation of other biological laws on different levels of the system in simulation models.

The conversion of raw starting materials into valuable products taking place in a biochemical system can be enclosed in a technological system of a bioreactor where most often living microbial cells represent the biological catalytic conversion device[2].

The degree of the overall complexity of a microbially catalyzed process is determined by the complexity of mutual relationships and interactions of the environment and the structured live matter while growing, utilizing and accumulating microbial metabolites. An appropriate mathematical description necessary for composition of a mathematical model has to respect the most important of these relationships and interactions. In case of complex systems, the systems analysis methodology recommends a break-down of these systems into individual sub-systems interconnected by well defined relationships which for microbial systems are usually determined by mass and energy transfer rates between individual sub-systems.

In contrast to technical systems where the subsystems are dimensionally comparable, in microbial processes this arrangement is hierarchical, thus, several subsystems on a certain hierarchical level make up a new subsystem on a higher hierarchical level. Figure 2.4 shows a possible alternative of breaking down a microbial process according to the hierarchical principle given by more or less natural boundaries. In the following paragraphs individual hierarchical levels, which could be distinguished in microbial systems, are briefly discussed.

Hierarchical Levels in Microbial Systems

(I)   The first hierarchical level is represented by subsystems concerning molecular or enzyme-catalyzed reactions. This group of subsystems includes all simple catabolic and anabolic reactions, reactions concerning material transport across the cell membrane, also synthesis and decomposition rates of macromolecules involved in catalytic activity, information transfer, energy storage, or those macromolecules having an important structural role. Connections among individual subsystems are determined by the reaction stoichiometry, mathematically usually expressed by the stoichiometric matrix of the reaction scheme. Mathematical models composed on this level, however, are very complex and rarely used. They are mainly encountered in basic research in the fields of biophysics, pharmacokinetics and metabolic disorders. For formulation of mathematical models concerning fermentation processes it is usually assumed that a certain subsystem determined by a reaction sequence between one state and another can be described with adequate accuracy by simplified kinetics based on rigidity or stoichiometric relationships and on the principle of a "bottleneck". These are the reasons why models of fermentation processes are usually derived from subsystems on a second hierarchical level.

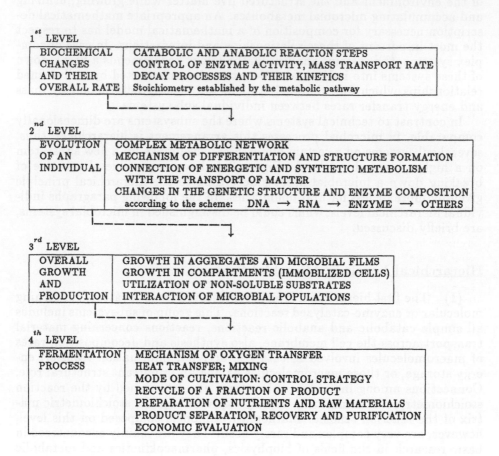

st **1 LEVEL**

| BIOCHEMICAL CHANGES AND THEIR OVERALL RATE | CATABOLIC AND ANABOLIC REACTION STEPS
CONTROL OF ENZYME ACTIVITY, MASS TRANSPORT RATE
DECAY PROCESSES AND THEIR KINETICS
Stoichiometry established by the metabolic pathway |

2nd LEVEL

| EVOLUTION OF AN INDIVIDUAL | COMPLEX METABOLIC NETWORK
MECHANISM OF DIFFERENTIATION AND STRUCTURE FORMATION
CONNECTION OF ENERGETIC AND SYNTHETIC METABOLISM
    WITH THE TRANSPORT OF MATTER
CHANGES IN THE GENETIC STRUCTURE AND ENZYME COMPOSITION
according to the scheme: DNA → RNA → ENZYME → OTHERS |

3rd LEVEL

| OVERALL GROWTH AND PRODUCTION | GROWTH IN AGGREGATES AND MICROBIAL FILMS
GROWTH IN COMPARTMENTS (IMMOBILIZED CELLS)
UTILIZATION OF NON-SOLUBLE SUBSTRATES
INTERACTION OF MICROBIAL POPULATIONS |

4th LEVEL

| FERMENTATION PROCESS | MECHANISM OF OXYGEN TRANSFER
HEAT TRANSFER; MIXING
MODE OF CULTIVATION: CONTROL STRATEGY
RECYCLE OF A FRACTION OF PRODUCT
PREPARATION OF NUTRIENTS AND RAW MATERIALS
PRODUCT SEPARATION, RECOVERY AND PURIFICATION
ECONOMIC EVALUATION |

Figure 2.4: A scheme of a fermentation process break-down by a systems analysis method.

(II)   The second hierarchical level is characterized by individual parts of metabolism, such as glycolysis, proteosynthesis, substrate transport, etc., being perceived as subsystems making up a complex which reacts to the outside perturbance by changing the rates of growth, substrate utilization and product formation. Models of microbial growth on this level may be represented by either one or several metabolic subsystems. Harder and Roels[25] in their review addressed the question of usefulness of dividing the growth model into a certain number of subsystems. They showed that the number of subsystems in the model depends on the length of relaxation times given by the rates of diffusion $(10^{-5} - 10^{-4}$ s$)$, of enzymatic reactions under allosteric control $(10^{-4} - 10$ s$)$, of RNA synthesis $(10 - 10^3$ s$)$, and also by the reaction times concerning the changes in enzymatic concentrations in the cell $(10^3 - 10^5$ s$)$. Considering that rates of diffusion fluxes are from the evolutionary standpoint, according to Snoll[74], counterbalanced by the enzymatic activity under allosteric control, it is possible to include these mechanisms, from the systems analysis standpoint, in one type of subsystems. The overall effect is that for a complete description of the dynamic response of the culture growth to outside perturbances it is quite adequate to use growth models consisting of three dynamically different systems[60].

The three dynamic subsystems differ in relaxation times, i.e., in the dynamics of simple reactions, in the dynamics of RNA synthesis, and in changes of enzymatic concentrations. The dynamic subsystem of RNA synthesis is particularly expressed in transient phases of fermentation, usually in a negative sense as exemplified by the lag phase following inoculation. This effect can be at least partially eliminated by an appropriate preparation of inoculum which is reflected in a model simplified to only one or two subsystems. A non-structural process description by a mathematical model can be used for shorter cultivations when the enzyme concentration can be considered as constant and the growth dynamics is practically dependent on the kinetics of simple reactions connected with catabolism. The non-structural description of the growth dynamics is the most frequent one even though it is not methodologically quite appropriate when used in the transfer of a process from a batch to a continuous-flow culture regime where the long-term changes in culture dynamics may be particularly pronounced in the elemental composition of microbial biomass as well as in enzyme concentrations and metabolic activity of the culture. The use of growth models segregated into several subsystems is, from the systems analysis point of view, an essential means for description of the dynamics of adaptation and selection having an influence on enzymatic composition and metabolic activity in cultivations with long retention times. When modeling cultivations with short retention times the most important aspect is represented by the requirement for simplicity of the mathematical description leading usually to a non-structured mathematical growth model. Apart from the growth dynamics, the model should also include the dynamics of the environment which is an equally important non-living part of the microbial culture.

(III)   The third hierarchical level in composing a model of a fermentation process is reflected in modeling of mutual relationships and links between or among microbial strains in mixed population cultivations of the predatory nature when one species serves as a "substrate" for another, or when two species on the same trophic level compete for the same substrate. Modeling of relationships between morphologically different individuals of the same species or those who differ in age also belongs in this category. It is also necessary at this stage to elucidate the effect of growth in colonies or aggregates on the overall growth rate. In practice, these problems are encountered when dealing with phenomena taking place in microbial colonies, microbial films or other natural or artificial aggregates such as cells immobilized in gels or pellets produced by higher microorganisms. These aspects are described in more detail by Ramkrishna[57] or Atkinson[2]. The use of this modeling level is justified only in cases where the above mentioned phenomena affect the overall production rate as the case may be in reactors with immobilized cells, with biological wastewater treatment systems, or with production of certain metabolites in reactors containing purposely cultivated microbial pellets. The segregated models, when compared to non-segregated ones, are considerably more complex and more difficult to solve which makes their use in fermentation technology rather limited to cases when the level of production depends significantly on the age distribution of individuals comprising the culture. These types of models are very significant, however, in the case of systems analysis applied to complex systems because they enable a sensitive simulation of conditions for ecological equilibrium and stability of microbial subsystems in natural environments. These problems are closely related with applications of the fourth level of modeling which then also incorporates system dynamics and properties of the environment.

(IV)   The fourth hierarchical level in modeling microbial system is characterized by linking the overall microbial growth and production rates with the dynamic balance of the environment distributed in space with all the attributes of the microbial environment such as kinetics of mixing, heat and mass transfer together with boundary conditions characterizing the intensity of mass and energy exchange with other subsystems, i.e. operations comprising the entire fermentation production process including medium preparation and product recovery.

When formulating a model of a microbial process, feasibility is a guiding principle. A very frequent mistake committed by those with little experience in this field is creation of a very complex model including different approaches available in the literature, disregarding their relevance to the overall goal which should always be the simplest and yet adequately accurate way of describing the real process which would enable its simulation by calculations. Such a model can then be conveniently used for the prediction of optimal operating conditions of a technological process as a whole.

2.1   KINETICS OF SIMPLE PROCESSES

The process of creating the mathematical model of a fermentation starts usually from a simplified scheme of reactions derived from a knowledge of metabolic pathways involved. Each metabolic reaction step is characterized by the reaction stoichiometry on one hand and by the flux, represented by the reaction velocity or rate, on the other. When composing the mathematical model, reaction rates are usually approximated by using one of the relationships derived from the theory of enzymic or chemical reactions. Table 2.1 summarizes the most frequently employed relationships suitable for describing the dynamics of individual metabolic sub-systems.

Clarification of the function and the physico-chemical meaning of individual mathematical reaction rate models listed in Table 2.1 is presented in the following paragraphs.

TABLE 2.1

**SUMMARY OF THE MOST FREQUENTLY USED RELATIONSHIPS
FOR EXPRESSING SIMPLE KINETICS IN THE SIMULATION
MODELS OF FERMENTATION PROCESSES**

$$r_1 = k\,S$$

$$r_2 = k\,S^n$$

$$r_3 = \frac{kS}{K+S}$$

$$r_4 = \frac{kS^n}{K+S^n}$$

$$r_5 = k\,[1 - exp\,(-S\,/\,K)]$$

$$r_6 = k\,exp\,(-S\,/\,K)$$

$$r_7 = \frac{kK}{K+S}$$

$$r_8 = \frac{kK}{K+S^n}$$

Reaction rate of r_1 type is characterized by a linear relationship between the rate of the phenomenon and the reaction substrate concentration. The reaction characterized by this type of reaction rate is basically controlled by diffusion. A linear relationship between the sub-process rate and the substrate concentration is not typical for biological processes because the overall rates of enzymatically catalyzed reactions are usually lower than the rate of diffusion which thus could not become the rate controlling parameter. Linear relationships between the reaction rate and the substrate concentration, however, are typical for decay processes within the cell, and for the influence of atypical substances in the cell as is the case, for example, in pharmaco-kinetics when modeling the response of the biological system to different pharmaceuticals and medicinal compounds. A number of these substances often pass through the organism without a change and their excretion is most often controlled by the rate of diffusional processes involved.

Reaction rate of r_2 type is derived from the Freundlich adsorption isotherm and it is typical for processes controlled by the physical adsorption onto solid surfaces or structures. This phenomenon is characteristic for most hydrolytic reactions. Kinetics of this type can be applied in the modeling of processes involving utilization of solid substrates such as cellulose, starch, etc.

Reaction rate of r_3 type is one of the most typical relationships used for modeling of fermentation processes because it represents the rate of change of a phenomenon controlled by chemisorption of the substrate onto one active site such as the molecule of an enzyme. When the rate constant k is of a very low value, this rate can be interpreted from the systems theory standpoint as a final automation functioning as an on-off switch of a subsystem with regard to the signal by the substrate concentration S. This relationship is used for the modeling of all situations where the model is to be employed in the simulation of a state characterized by phasing off of the metabolic subsystem when a certain substance runs out and also vice versa. The value of constant k then represents the maximum velocity with which the given subsystem can participate in the overall process.

Reaction rate of r_4 type is a modification of the previous case of chemisorption where more than one active site is present on each biocatalytic molecule. Kinetics of this type is useful for describing the dynamics of the reaction mechanism on multienzymatic complexes, specialized organelles or, as reviewed by Harder and Roels[25], when modeling the synthesis of catabolic enzymes which is controlled by the rate of mRNA synthesis in turn dependent on chemisorption of CA protein and cAMP on the promoter site of chromosomal DNA. The use of this method of describing the reaction rate of a sub-process, however, is not very frequent in the modeling practice because the resulting equation contains parameters difficult to identify, and because sigmoidal behaviour of the key sub-systems in biosynthesis is not very usual. The subsequent process simulation calculations may be significantly slowed down when this type of model with non-integer-number values of parameter n is used because numerical expression of quantity S^n is done by a double use of a series for numerically expressing the exponential and the logarithm.

Reaction rate of r_5 type is a somewhat unusual type of rate relationship suggested for describing the process dynamics. It is based on a purely physical interpretation derived from equations for movement of a mass point in a surrounding characterized by dissipation forces. The force exerted by a point of unit mass against the resistance of its surroundings is expressed by a basic equation

$$\dot{v} = k \cdot v^2$$

Upon substitution for velocity v defined in the state space by the concentration whereby

$$v = r = -\frac{\mathrm{d}S}{\mathrm{d}t} = \frac{\mathrm{d}P}{\mathrm{d}t}$$

and following a rearrangement of the original equation where time is eliminated, the modified equation can be solved for boundary conditions on a trajectory where

$$(S = 0 ; r = 0) \quad and \quad (S \gg 0 ; r = k)$$

The use of r_5 type of kinetics is not so frequent in composing a mathematical model for process simulation. It could be used, however, for the description of sub-process rates with regard to the internal limits of the reacting system as the case is, for example, in modeling reactions concerning the incorporation of trace elements or phosphate in various cell structures or in the cell wall.

Reaction rate of r_6 type is derived from a similar physical hypothesis as for the above kinetic model. The substance with concentration S, however, is considered as directly participating in the dissipation of kinetic energy during the course of the reaction. The relationship for the dissipation of kinetic energy, however, is solved for different boundary conditions on the metabolic trajectory $(S = 0; r = k)$. The practical use of this equation is in the simulation of a reaction slow-down (reaction of the zeroth order) and thus for expressing some feedback in the process. This particular type of kinetics, similarly to r_5 type, is not extensively used in practice. Its physical interpretation may be, when compared to the inhibition model derived from enzyme kinetics, somewhat uncertain.

Reaction rate of r_7 type is based on the principle of hypothetical reversible blocking of the active reaction site by chemisorption of a substance with concentration S. This substance blocks the enzyme activity and causes an effect demonstrated by negative inhibition. Models of this reaction rate type can be used in the process simulation model as terms capable of realizing a negative feedback in the metabolic scheme of the process.

Reaction rate of r_8 type is a variant of the above (r_6) type derived for inhibition of a larger number of active reaction sites of a certain biochemical process bottleneck. An advantage of this approach, which uses higher values of exponent n, is that it enables simulation of a complete switch-off of the metabolic sub-system when the substance concentration S exceeds certain final value. This is accomplished through a sigmoidal-type function. Identification and computing problems associated with this type of simple kinetics are similar to those mentioned in connection with r_4 type above.

When constructing a mathematical model of a fermentation process, the use of simple rate relationships listed in Table 2.1 would cover the need for most cases addressed on the first or second level of system hierarchy. Very often the use is made of combinations of these rate relationships based on superimposition of several simple phenomena in the given sub-system. The mode of superimposition depends on mutual relationships among the sub-systems. Most often the summation or multiplication product is used depending on whether there is a sequential or alternative effect according to the following relationships:

$$r = \sum_{i=1}^{n} r_i \tag{2.1}$$

$$r = \prod_{i=1}^{n} r_i \tag{2.2}$$

In the following section a few examples will illustrate how to formulate the kinetic relationship (equation) based on observed experimental data for a given process.

EXAMPLE 2.1

When cultivating the yeast *Candida boidinii* on methanol it has been observed that for low methanol concentrations the process is limited by the substrate concentration while at high concentrations the specific growth rate was lowered. A mathematical relationship for expressing the specific culture growth rate is to be suggested.

Solution:

This is the case of alternating effects and for the process simulation based on a mathematical model, a combination of relationships of a type r_3 and r_4 should be based on Equation (2.2). The resulting relationship for the specific growth rate will then assume the following form:

$$\mu = k \frac{S}{K_1 + S} \frac{K_2}{K_2 + S}$$

Following an algebraic rearrangement, the relationship could assume a well-known form:

$$\mu = \mu_{max} = \frac{S}{K_S + S(1 + \frac{S}{K_I})}$$

whereby K_S and K_I are the saturation and inhibition constants, respectively.

EXAMPLE 2.2

The yeast *Saccharomyces cerevisiae* grown aerobically on glucose utilizes the sugar and while the culture grows it also produces some ethanol. When the sugar is depleted the culture starts to consume the alcohol. An appropriate expression for the specific culture growth rate is sought.

Solution:

Glucose (S_1) and subsequently ethanol (S_2) are the culture growth limiting carbon-containing substrates which are utilized in a sequential manner.

Glucose inhibits the growth when ethanol serves as a growth substrate. For the utilization of glucose the reaction rate of r_3 type can be used. The ethanol utilization can be described by a combination of r_3 and r_7 reaction rate types. The resulting expression for the specific growth rate can then be as follows:

$$\mu = \underbrace{k_1 \frac{S_1}{S_1 + K_1}}_{\text{growth on glucose}} + \underbrace{k_2 \frac{S_2}{S_2 + K_2} \frac{K_3}{K_3 + S_1}}_{\text{growth on ethanol}}$$

This model can for instance be applied for simulation of a catabolic repression in the process of baker's yeast production.

EXAMPLE 2.3

During the anaerobic cultivation of *Clostridium acetobutylicum* bacteria it was observed that the growth rate is inhibited by accumulating butanol (P) and limited by the decreasing sugar concentration (S). Following the depletion of sugar, the cell lysis was observed which was proportional to the butanol product concentration. Formulation for the specific growth rate is to be devised.

Solution:

There are three effects in this cultivation whereby the limitation and inhibition are alternating phenomena while the lysis is sequential. For the construction of the model suitable for the simulation of the culture growth behaviour the combination of reaction rates of types r_1, r_3 and r_7 can be used giving the following relationship for the specific growth rate:

$$\mu = k_1 \frac{S}{S + K_1} \frac{K_2}{K_2 + P} - k_2 P$$

A model of this type was used for evaluation of experimental data from a small pilot plant. The model was capable of reflecting well the culture experiment in its exponential, stationary and declining phases. The use of this model also for the culture lag phase will be shown in Section 4.3.

2.2   STOICHIOMETRY OF MICROBIAL PROCESSES

While kinetic relationships characterize the dynamic properties of individual parameters of the fermentation process, stoichiometry determines mutual relationships and internal limitations within the biochemical system. For a truly chemical process, where the composition of all the components is well known, the stoichiometry can be mathematically described for instance by the stoichiometric matrix of the process[1]. The matrix expression of stoichiometric relationships is very useful when mathematical models are formulated for the given process and particularly when computers are used for process simulation studies. For microbial systems this expression could be used when formulating the model on the first level of systems analysis[60]. This approach, however, would require a detailed knowledge of metabolic pathways and compositions of all components involved in the process studied. Such may be the case concerning some isolated subsystems under *in vitro* conditions or when the main metabolic pathways are studied.

The conventional chemical way of solving stoichiometric problems given by the relationships among sub-systems is too cumbersome for most cases of complex biological systems and cannot be conveniently used. With models on higher hierarchic levels of the process systems analysis the relationships among sub-systems could be expressed by macroscopic stoichiometric coefficients respecting the laws of mass and energy conservation. More detailed methodology for determination of macroscopic coefficients can be found in the original papers of Minkevich and Eroshin[48], Erickson *et al.*[14], and in the review by Roels[64].

The macroscopic mass balance of the sub-system concerning the biomass production and another product can be written in its general form[14]:

$$CH_mO_l + aNH_3 + bO_2 = y_c\ CH_pO_nN_q + z\ CH_rO_sN_t + (1 - y_c - z)\ CO_2 + cH_2O$$
$$(Substrate) \qquad\qquad (Biomass) \qquad (Product)$$

Compositions of the substrate, biomass and of the product in this equation are expressed by the elemental chemical analysis. The biomass composition can be approximated by a generally valid formula[64] $CH_{1.79}O_{0.5}N_{0.2}$ with the error usually smaller than 5%. While writing the mass balances for the sub-system it is necessary to respect the elemental composition of the biomass, substrate and product and also the degree of reduction which expresses the number of free electrons for 1 gramatom of carbon and which is described by the following relationships for individual components:

$$Biomass: \qquad \gamma_B = 4 + p - 2n - 3q \qquad (2.3a)$$
$$Product: \qquad \gamma_P = 4 + r - 2s - 3t \qquad (2.3b)$$
$$Substrate: \qquad \gamma_S = 4 + m - 2l \qquad (2.3c)$$

where the number of free electrons is taken as 4 for one atom of carbon, 1 for the atom of hydrogen, -2 for the atom of oxygen, and -3 for the atom of nitrogen. Based on this consideration it is obvious that in the metabolic end-products such as H_2O, CO_2 and NH_3 there is no free electron, while oxygen in the O_2 form can accept 4 electrons. The free electron balance, when the macroscopic chemical equation of the process is considered, will have the following form:

$$\gamma_S + b\,(-4) = y_c\,\gamma_B + z\,\gamma_P \tag{2.4}$$

The oxygen demand can then be expressed as:

$$b = (\gamma_S - y_c\,\gamma_B - z\,\gamma_P)\,/\,4 \tag{2.5}$$

When Equation (2.4) is divided by γ_S, a relationship results which reflects the free electron content of the organic substrate:

$$\frac{4b}{\gamma_S} + \frac{y_c\,\gamma_B}{\gamma_S} + \frac{z\,\gamma_P}{\gamma_S} = 1 \tag{2.6}$$

The first group represents the free electrons taken away from the substrate by oxygen, the second one reflects the fraction of substrate energy transferred into the biomass, and the third group reflects the fraction of substrate energy represented by free electrons in the product. The first group also includes the fermentation heat which, expressed for the unit organic substrate containing 1 gramatom of carbon, can be expressed according to Minkevich[48]:

$$Q = 4Q_o\,b \quad [\,kJ\,/(\,1\;gatom\;of\;C\,)\,] \tag{2.7}$$

where Q_o is approximately 113 kJ *equivalent of free electrons* transferred from the substrate to CO_2. The invariant Q_o directly links the mass balance of the process with its energy balance.

The second group in Equation (2.6) represents the energy yield coefficient for biomass production

$$\eta = y_c\,\frac{\gamma_B}{\gamma_S} \tag{2.8}$$

This coefficient expresses the ratio between the biomass combustion heat and that of the substrate when NH_3 is present as a nitrogen source. When another nitrogen source is used in the process the relationship can be modified (as recommended by Roels[64]).

The third group in Equation (2.6) represents the fraction of substrate energy found in the product and can be expressed as:

$$\xi_P = z \; \frac{\gamma_P}{\gamma_S} \tag{2.9}$$

Another way of characterizing compounds participating in the sub-process is to use a parameter introduced by Minkevich *et al.*[14,48] which reflects the weight fraction of carbon in organic matter defined by the following relationships:

$$\text{Biomass}: \qquad \sigma_B = 12/(12 + 2p + 16n + 14q) \tag{2.10a}$$
$$\text{Product}: \qquad \sigma_P = 12/(12 + 2r + 16s + 14t) \tag{2.10b}$$
$$\text{Substrate}: \qquad \sigma_S = 12/(12 + 2m + 16l) \tag{2.10c}$$

The process efficiency can then be expressed with regard to the actual values of macroscopic yield coefficients $Y_{X/S}$ and $Y_{P/S}$ determined by experimental laboratory measurements:

$$Y_{X/S} = \eta \; \frac{\sigma_S \, \gamma_S}{\sigma_X \, \gamma_X} \tag{2.11}$$

$$Y_{P/S} = \xi_P \; \frac{\sigma_S \, \gamma_S}{\sigma_P \, \gamma_P} \tag{2.12}$$

and for oxygen

$$Y_{X/O} = \frac{3 \, \eta}{2 \, \sigma_B \, \gamma_B \, (1 - \eta - \xi_P)} \tag{2.13}$$

The process yield coefficient with regard to nitrogen can be similarly derived from the species mass balance for nitrogen which is given as

$$a = y_c \, q + z \, t \tag{2.14}$$

This balance enables an easy calculation of the inorganic nitrogen consumption for growth and production as will be shown in the following examples. For practical calculations it is important that the thermodynamic efficiency of aerobic processes is given by the sum of η and ξ_r, ranging between 0.55 and 0.6. It is approximately 0.7 for anaerobic processes. This fact can assist in the estimation of yield coefficients for new fermentation processes or for pointing out inadequate or erroneous measurements of the process studied.

EXAMPLE 2.4

Elemental analysis of refuse revealed that it contained 76%(w/v) of organic matter of the composition $CH_{2.1}O_{0.9}N_{0.15}$.

The remaining 24% was ash. The maximum methane production by anaerobic digestion is to be assessed calculated per 1 kg of dry refuse.

Solution:

The overall stoichiometric equation for the methanogenic process is:

$$CH_{2.1}O_{0.9}N_{0.15} + aH_2O \rightarrow yCO_2 + xNH_3 + zCH_4 + bH^+$$

If no other information is available on the process thermodynamic efficiency, it can be assumed that the maximum value for $\eta_P = 0.7$. Upon application of Equation (2.12) the maximum process yield can be estimated whereby the individual quantities required for the calculation can be calculated as follows:

$$
\begin{aligned}
\sigma_s &= 12/(12 + 2.1 + 16 \times 0.9 + 14 \times 0.15) &= 0.392 \\
\sigma_P &= 12/(12 + 4) &= 0.75 \\
\gamma_s &= 4 + 2.1 - 2 \times 0.8 - 3 \times 0.15 &= 4.05 \\
\gamma_P &= 4 + 4 &= 8
\end{aligned}
$$

The yield coefficient based on dry substrate with ash requires modification of σ_s whereby the following relationship is used

$$\sigma'_s = \sigma_s (1 - 0.24) = 0.29$$

The maximum yield coefficient is calculated by using Equation (2.12):

$$Y_{CH_4/s} = 0.7 \frac{0.298 \times 4.05}{0.75 \times 8} = 0.141 \text{ kg } CH_4/\text{kg dry refuse}$$

By using Equation (2.9) the value of stoichiometric coefficient z can be determined

$$z = \frac{1}{\xi_P} \frac{\gamma_s}{\gamma_P} = \frac{1}{0.7} \frac{4.05}{8} = 0.72$$

This value is important because it determines the maximum percentage of methane in the biogas, complemented by 28% (v) CO_2, because the CO_2 content has to conform to the carbon mass balance:

$$1 = y + z \qquad (y = 0.28)$$

The nitrogen balance indicates that the value of stoichiometric coefficient x is:

$$x = 0.15$$

From the mass balances for oxygen and hydrogen the value of coefficient a can be derived:

$$a = 2y - 0.9 = 0.54$$

$$2.1 + 2a = 3x + 42 + b$$

then $\qquad\qquad b = 0.15$

This numerically determines the mass balance of the process.

EXAMPLE 2.5

The following macroscopic coefficients should be determined for the cultivation of the yeast *Candida utilis* on ethanol:
- substrate consumption for biomass
- oxygen consumption
- fermentation heat evolution

Solution:

The overall process chemical stoichiometry can be recorded in the manner suggested by Erickson[14]:

$$CH_3O_{0.5} + bO_2 + aNH_3 = yCH_{1.79}O_{0.5}N_{0.2} + xCO_2 + zH_2O$$
$$(ethanol) \qquad\qquad\qquad (biomass)$$

Since biomass is the only product of this process with a non-zero internal energy content, Equation (2.12) can be used for estimation of the process yield coefficient whereby the maximum value for anaerobic processes would be $\eta = 0.6$. Equations (2.3) and (2.10) will be used for calculations of values for $\sigma_x, \sigma_s, \gamma_x$ and γ_s. The yield coefficient then can be determined by substituting into Equation (2.11):

$$Y_{x/s} = \eta \frac{\sigma_s \gamma_s}{\sigma_x \gamma_x} = 0.6 \frac{(0.522)\ 6}{(0.488)\ 4.19} = 0.92 \text{ (kg biomass / kg substrate)}$$

This value is very close to the actual values quoted e.g. by Minkevich[48] or Roels[64].

The oxygen-based yield coefficient for this process can be determined by using Equation (2.13) for $\eta_P = 0$.

$$Y_{X/O} = \frac{3\ (0.6)}{2\ (0.488)\ 4.19\ (1 - 0.6)} = 1.1 \ (\text{ kg biomass / kg oxygen used })$$

Stoichiometric coefficient b is required for determination of the heat of fermentation. It can be expressed by rearrangement of Equation (2.6):

$$b = (\gamma_S - \eta\ \gamma_S)/4 = 0.6$$

Fermentation heat can then be calculated by using Equation (2.7)

$$Q = 4 \times 133 \times b = 319.2 \ (\text{kJ / gramatom C})$$

Upon converting this result onto the substrate basis we obtain

$$Q' = Q\ \sigma_S\ /\ 12 = 13.8 \ (\text{ kJ / g substrate})$$

From the nitrogen balance the value of coefficient a can easily be determined

$$a = 0.2\ y = \eta \frac{\gamma_S}{\gamma_B}\ (0.2) = 0.172$$

thus enabling estimation of ammonia consumption in the process.

$$Y_{NH_3/X} = (0.2)\frac{17}{12}\ \sigma_X = 0.14 \ (\text{kg NH}_3 \text{ / kg biomass})$$

This calculated value is very close to the experimentally determined values for the aerobic culture production of fodder yeast from the synthetic ethanol substrate.

The two examples above illustrate the usefulness of the macroscopic balancing of fermentation processes. Both examples were of a simple character because more complex processes with several products in addition to just biomass would require a more sophisticated approach beyond the scope of this brief Chapter. The subject is dealt with in more detail in the published material cited[14,48,64].

2.3 PHYSIOLOGICAL ASPECTS OF MATHEMATICAL MODELS FOR FERMENTATION PROCESSES

The way mathematical models of fermentation processes have been formulated has traditionally been influenced by the chemical reactor theory and by the results of studies in the field of enzyme kinetics. The behaviour of a microbial culture is often being simplistically perceived as that of a chemical reactor whereby the long-term changes based on the variable nature of biological activity are neglected. Such an approach may be adequate when modeling fast-growing microbial cultures on a defined medium where the terminal product is the biomass. For long-lasting cultivations and/or production of secondary metabolites it is often obvious that describing the culture system as a simple chemically-based one represents only a very crude approximation of the reality. As was explained in the opening Chapter, under the prolonged-time culture conditions the effects may become apparent of changes in the enzymatic system and in the nucleic acid concentration of the culture which are related to its growth rate. Microbial physiologists very often criticize the inability of mathematical models developed for the description of microbial behaviour to describe and quantitatively express the culture history which affects the subsequent productivity of the culture. This relationship between the culture *history* and its productivity in the process, which is often demonstrated by changes even in the culture morphology, was often used to explain the controversial behaviour of a culture characterized by the same growth rate under the conditions of either batch or continuous culture. The concept of the physiological state of a microbial culture was introduced during discussions concerning the principles of the continuous culture in the late 1950's and the formulation of this concept has been gradually improving ever since. In 1974 Malek[45] formulated the following revised version of what is perceived as "the physiological state of a microbial culture".

> *"The term physiological state of microbial populations is an auxiliary operational term including their expressed properties and states as well as their latent potentialities which, depending on conditions in the medium, are reflected in a certain mode of reproduction and in the ability to respond to changes in these conditions by specific changes in the metabolism thereby creating a new quality, either in the form of a changed morphology (filaments, spores, etc.) or in the form of altered physiological and biochemical processes, often accompanied by formation and accumulation of specific metabolites."*

From the viewpoint of systems analysis applied to microbial processes, this definition covers most of the basic axioms derived from the hierarchical organization of the microbial process dynamics particularly with regard to slow changes given by adaptive and selective mechanisms which affect the overall biological activity of the microbial culture.

In the following paragraphs an attempt will be made to discuss the possibility of a quantitative description of individual phenomena derived from the formulation of the physiological state as it pertains to the construction of mathematical models for fermentation processes.

Latent potentials depending on culture conditions

The latent potentialities of microorganisms are coded in chromosomes and in other nonchromosomal DNA structures (like mitochondria, plasmids, etc.) Obviously, under "normal" conditions these potentialities are repressed or activated by excess of intracellular energy (in the form of ATP and cAMP), by substrates, by intermediate metabolites or by other physical factors. When the repression is removed by technological manipulation, the production of unexpected amount of different chemical species may take place. An example is the production of chlortetracycline of *Streptomyces aureus* when phosphate repression is removed. Another example is the excessive production of exocellular enzymes such as α-amylase, proteinase, lipase, etc. when repression by amino acids in the media is avoided. The production of toxins by *Penicillium chrysogenum* at lower temperatures (4° C) or the production of ethyleneglycol by fodder yeast *Candida utilis* when the pH of broth is kept at pH higher than 10 can serve as further examples. To control the biotechnological process on the basis of this it is necessary to compose mass balances for the key process components and the control is accomplished usually by means of changes in nutrient media composition and/or by feeding of chemical precursor compounds to push the physiological functions to the state of excessive production or overproduction of certain intracellular or extracellular metabolites.

The latent potentials of the culture are demonstrated by changes in metabolic activities and can be described by the variability in the composition of biomass. Based on this finding Frederickson[16] derived a concept of the vector physiological state particularly as a means for formulation of segregated structured models[25,57]. For this formulation it is assumed that the physiological state of the cell is described by a vector with a finite dimension $Z = (Z_1, Z_2, ... Z_n)$ which is represented by individual components Z_i comprising the biomass, while the environment is characterized by the finite concentration vector $C = (C_1, C_2, ... C_m)$ consisting of concentrations of individual compounds participating in biological reactions. The growth rate is then given by vector $\dot{Z} = (\dot{Z}_1, \dot{Z}_2, ... \dot{Z}_n)$ representing the rate of synthesis of individual cellular components. From a formal standpoint this concept is correct, however, upon practical application it does not explicitly include certain limitations in biosynthesis of individual components which are determined by internal stoichiometric relationships and by the hierarchy of their origin based on the internal economy of the process whereby the largest change in the composition acts against the change in the environment[74].

During this process the maximum biosynthetic rate is maintained. This applies also to a constant ratio of the content of water and dry matter as well as for the proportions of chemical elements which are based on the macroscopic mass balances. Also, the overall thermodynamic efficiency of the process is maintained within a relatively narrow interval as discussed in the preceding paragraph. Details of these aspects were presented in a related work by Ricica and Votruba[60]. In this paper the application of systems theory (state space methodology) was discussed from the point of view of modeling of the physiological state of microbial culture. The basic methodology of model building can be deduced from the shape of trajectories in the metabolic field. The model of a microbial culture as a chemical reactor is very reductive because these models are formulated on the conception that the culture enzyme activity is fixed as a catalyst concentration. Therefore, the physiological model has to include the dynamic balances of enzyme activity and their cooperation, like participants in a complex game with mass constraints.

For practical process simulations on a computer the Frederickson method[16], based on a vector concept of describing changes in the physiological state of the culture, proves to be somewhat complex and cumbersome because it requires calculations for a large number of cellular mass components and the resulting formulation is very confusing. For description of certain phenomena concerning transitional culture states or those encountered when modeling inoculum quality it is useful to apply the concept developed by Powell[56] based on his studies of bacterial culture age theory.

The variability of metabolic activity related to the culture history
can be described according to Powell[56]. In his theoretical work dealing with modeling of transient states in the microbial culture he introduced an interesting possibility of modeling the relationship between the specific growth rate μ, the culture history and the variable environment using the "metabolic activity functional" $Q(t)$ which is defined as a mean metabolic activity of the population. It is not a simple function of time but its value depends on the culture history and the variable substrate consumption rate during different stages of development of the microbial population. This functional has been defined as

$$Q(t) = \int\limits_{0}^{\infty} f(\xi) \, q \, [S(t-\xi)] \, \mathrm{d}\xi \tag{2.15}$$

where $f(\xi)$ is a distribution function representing the age structure of the population, and $q(S)$ is the metabolic activity of an individual cell depending on the substrate concentration S. The function $q(S)$ is a simple function depending on the environment; in a very simplified case it is equal to unity when there is some metabolizable substrate present in the culture broth or it becomes zero when there is none. Even though the $Q(t)$ metabolic functional has been formally derived from the variable morphological culture image which depends on the culture age, according to Powell[56] it can be considered identical with a variation of a concentration of some cell component which is related to the growth rate[16]. This component is then used as the "marker of the culture physiological state". Powell[56] also connected the metabolic activity functional with the macroscopic growth balance by the following equation for the specific growth rate μ:

$$\mu = Y_{s/x}\, r_s\, Q(t) \tag{2.16}$$

where $Y_{s/x}$ is the theoretical yield coefficient determined in Equation (2.11) for $\eta = 1$, and r_s is the specific rate of the limiting substrate utilization. This way the relationship between the thermodynamic efficiency η and the metabolic activity of the culture can be expressed:

$$\eta = \frac{1}{\tau} \int_0^\tau Q(t)\, \mathrm{d}t \tag{2.17}$$

The metabolic activity functional $Q(t)$ includes the history of the culture quantified by the age distribution function $f(t)$ and simultaneously links the population balance with macroscopic principles of balancing the fermentation process. Powell[56] recommended that the metabolic activity functional should be related to a suitable marker of physiological state which can be analytically determined. A suitable marker of this kind could perhaps be the concentration of RNA in the culture which is linearly related to growth:

$$\mu = \lambda\, (\mathrm{RNA} - \mathrm{RNA}_{\min}) \tag{2.18}$$

For practical calculations, a dimensionless variable y can be introduced and defined in the following manner:

$$y = \frac{\mathrm{RNA}}{\mathrm{RNA}_{\min}} \tag{2.19}$$

where $\mathrm{RNA}_{\min}$ is the RNA concentration in the cell at $\mu = 0$. The dimensionless parameter y characterizes the inoculation initial conditions by the value of $y = 1$. Equation (2.18) can then be transcribed using the dimensionless RNA concentration and taking advantage of the findings published by Harder and Roels[25] who concluded that the relationship between the culture RNA concentration and its growth rate is approximately constant for most bacterial cultures:

$$\mu = 0.56\, (y - 1) \tag{2.20}$$

This relationship can easily be substituted into dimensionless differential mass balances written for the cellular RNA concentration. Through this particular mass balance, if it constitutes one of the mass balances characterizing the bioreactor system and its behaviour, the so devised mathematical model of such a system becomes structured reflecting, to a certain degree, the physiological state of the culture in the bioreactor.

Morphological culture changes,

mycelial type of growth and spore formation can be linked, according to Snoll[74], to evolutional processes controlled in the direction of the biosynthesis gradient in the state space of the microbial system. Enzymes bound onto morphological structures of the cell show much higher specific rates of biosynthesis than enzymatic systems freely dispersed in solution because with structured enzymatic systems the effects of diffusional limitations appear to be much lower. The structured and organized enzymatic system possesses a much higher chance of reproduction in the mutual competition for energy related to the substrate.

The relationship of morphological changes can be described mathematically by using the theory of structural stability published by Thom[82], and eventually using the theory of self-organization recently published by Haken[24]. The works of so called "Brussels school" of Prigogin and Nicolis[50] are also very relevant. A practical solution of three dimensional problems of structure formation and changes in morphology in multi-component reaction systems is a challenge for the future student which will require high-performance computers and effective problem-oriented software together with development of a theoretical apparatus suitable for elucidation and interpretation of experimental results in this field.

A book written by Thompson[84] is dedicated to the application of the "catastrophe" theory in science and engineering. It shows how a simple system, whose dynamic behaviour is described by ordinary differential equations, can simulate nonsmooth behaviour and jump from one state to another in a discrete way. Such phenomena are typical when the cell is divided into two parts. The continuous biochemical processes can, under certain conditions produce the branching points at which a little signal can drive the differentiation process to one or another region independently of genetic information but dependent on the conditions in the environment. In the population dynamics the classical predator-prey model is analyzed and such situation can be of importance when contamination in an industrial fermentation tank occurs.

EXAMPLE 2.6

A mathematical model is to be constructed which would simulate the growth of a microbial culture during the transitional phase from inoculation which was done by using an inoculum in a stationary phase ($\mu = 0$) to an exponential phase of growth ($\mu = \mu_{max}$) by using the intracellular concentration of RNA as a marker of physiological state.

Solution:

The relationship between the RNA concentration and the growth rate can be expressed

$$(\text{RNA}) = a + b\,\mu$$

When describing the growth phenomenon it can be assumed that the overall RNA synthesis rate is proportional to the amount of RNA already present in the culture:

$$\frac{d(\text{RNA})}{dt}\, X = k\, X\, (\text{RNA})$$

Upon differentiation of this relationship and subsequently dividing it by X (biomass concentration) an equation results describing the dynamics of the RNA synthesis as related to the concentration of RNA and biomass:

$$\frac{d(\text{RNA})}{dt} = \left(k - \frac{1}{X}\frac{dX}{dt}\right)(\text{RNA})$$

This mass balance relationship is supplemented by the one for biomass

$$\frac{dX}{dt} = \mu X = \frac{(\text{RNA}) - a}{b}\, X$$

When the initial condition $t = 0$, $X = X_o$, RNA $= a$, ($\mu = 0$) is substituted, the two differential equations above can be solved e.g. numerically. It is found that by using this approach the overall phenomena taking place during the lag phase can be expressed and modelled. Figure 2.5 illustrates the results of the numerical solution of the two differential equations above for $X_o = 0.1$ and values of parameters $a = 0.08$ and $b = 0.16$ derived for the relationship of the weight ratio of RNA with the growth rate for different types of bacteria[25]. The value of constant $k = 0.5$ which is the value of μ_{max} as seen from the steady-state value for RNA.

From this example it is seen that even physiological observations can be interpreted in a mathematical form which can be used for process simulation using a computer-based approach.

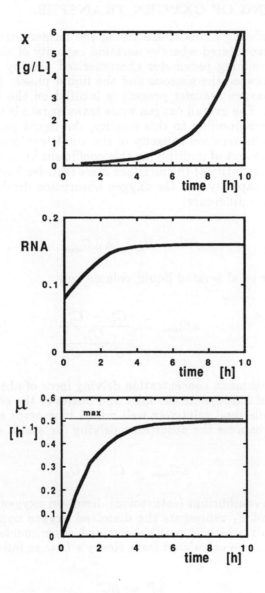

Figure 2.5: Modeling of the culture lag phase by using RNA as a marker of the culture physiological state. Results of the numerical solution of two differential equations describing the fermentation process in Example 2.6.

2.4 MODELING OF OXYGEN TRANSFER

When formulating a model describing the fermentation process, a case is very often encountered whereby aeration capacity of the system appears to be the determining parameter characterized also by the oxygen mass transfer rate between the gaseous and the liquid phase. The highest resistance for the oxygen transfer process is located on the side of the liquid phase interface. The overall oxygen mass transfer rate is thus controlled by a coefficient corresponding to this locality, the liquid phase mass transfer coefficient k_L. The aeration capacity of the whole equipment is then determined by the product of multiplying this coefficient by the specific interface surface area a. Quantity of the interface mass transfer flux can be expressed by the aeration capacity and the oxygen absorption driving force based on the concentration difference:

$$N = V_R \, k_L \, a \, \Delta C_{mean} \tag{2.21}$$

where V_R is the total aerated liquid volume and

$$\Delta C_{mean} = \frac{C_1^* - C_2^*}{\ln \dfrac{C_{L1}^*}{C_1^* - C_{L2}^*}} \tag{2.22}$$

is the logarithmic mean concentration driving force of absorption. Numerical indices 1 and 2 designate the inlet and outlet of the equipment, respectively. For smaller and relatively well mixed bioreactor equipment it may be adequate to express the absorption driving force as a simple difference

$$\Delta C_{mean} = C^* - C_L \tag{2.23}$$

where C^* is the equilibrium (saturation) dissolved oxygen concentration at the interface and C_L represents the dissolved oxygen concentration in the liquid broth bulk which is best determined experimentally. The former quantity (C^*) can be calculated from Henry's Law as follows:

$$yP = HC^* \tag{2.24}$$

where:
  y = mole fraction of oxygen in the aeration gas
  P = gas pressure
  H = Henry's Law constant for oxygen

The product of yP represents the partial pressure of oxygen.

The formulation of simulation models for fermentation processes depends a great deal on the fermentation conditions. When the microbial process takes place in an intensively aerated bioreactor with a lower biomass concentration it may be adequate for the mass balance calculations to combine the oxygen transfer rate $k_L a$ with the oxygen utilization rate by the microorganisms. The model using this approach as a basis is likely to represent the actual behaviour of the bioreactor reasonably well. When formulating the model, a mass transfer scheme is assumed whereby the first step is based on the physical gas-liquid transfer, followed by the dissolved oxygen consumption by the microorganisms. For describing this phenomenon the rule of reaction velocity summation, as specified by Equation (2.1), can be applied for the bioreactor subsystem.

In some extreme cases, when the aeration rate in comparison with the microbial respiration potential is low, the oxygen limitation of the process occurs and in this case, apart from the gas-liquid-microbes transfer mechanism, a shortened mass transfer alternative may be established whereby oxygen is directly utilized by microorganisms adsorbed on the gas-liquid interface.

The topic of the physical mass transfer controlled by the kinetics of microbial flotation has been elucidated in some detail in papers published by Sobotka et al[75,76]. The direct oxygen transfer has to be accounted for in the overall oxygen mass balance, particularly for fermentation media containing alcohols or other surface active substances. Dimensions of microbial cells are also important because the interphase forces can more easily keep smaller cells at the interface. A review paper on the modeling of oxygen transfer in aerobic microbial cultures by Sobotka et al.[77] presents a useful summary of this topic.

EXAMPLE 2.7

Formulate a model for oxygen transfer for a batch cultivation of *Candida boidinii* yeast on methanol. The culture is propagated in a well aerated laboratory-size fermentor.

Solution:

Considering the methanol toxicity (see Example 2.1), it can be assumed that the working concentration of methanol in the broth will be low. This fact justifies the use of a simpler mass transfer model without the direct oxygen transfer to the growing cells. The laboratory fermentor assures close-to-ideal mixing conditions, the mass balance in the liquid phase becoming then:

$$V_L \frac{dC_L}{dt} = V_L \, k_L a \left(\frac{P}{H} y - C_L \right) - V_L \, Y_{O/X} \, \mu X$$

$$(accumulation) = (interphase\ transfer) - (respiration\ oxygen\ consumption)$$

where:

C_L	=	dissolved oxygen concentration in the liquid broth
P	=	total gas pressure
H	=	Henry's constant for oxygen
$k_L a$	=	aeration capacity of the bioreactor
V_L	=	volume of liquid broth
y	=	volume fraction of oxygen in the gas phase
$Y_{O/X}$	=	stoichiometric coefficient of oxygen consumption for growth (see Equation 2.13)
μX	=	overall biomass growth rate.

2.5   THE USE OF SIMPLE MIXING MODELS FOR SIMULATION OF FERMENTATION PROCESSES

When modeling a real system where the relationship between the individual cell growth rate and the whole population growth rate as well as the relationship of this population to the environment are to be expressed, the variability of quantities characterizing the environment has to be respected. In batch processes the latter is a function of time, but for the large-scale process it is also determined by the location of individuals (individual cells e.g.), and substrate (and/or metabolite) concentration gradients in space determined by the macroscopic mass balances on the equipment.

The transfer of results, or scale-up, from close-to-ideal laboratory conditions where there it is possible to maintain a concentration homogeneity within the whole cultivation space to a larger-scale process, has been known to present a problem. It is extremely difficult to realize conditions of ideal mixing in large-volume bioreactors particularly when viscous media are being used or culture conditions develop characterized by non-Newtonian broth behaviour. The conditions of non-ideal mixing can be artificially "simulated" by cell immobilization. A complete mathematical description of the system based on rigorous application of the mass and momentum principle would be possible for a multiphase, multicomponent system. However, the use of this formal apparatus for process simulation calculations could be very complex and cumbersome. Therefore, for simplicity of the model system used to describe the process, the use is being made for simulation of non-ideal mixing systems of either models combined from individual ideal subsystems (compartmentalized models) or very simplified forms of phenomenological equations (quasi-homogeneous models with backmixing or dispersion models for short). This approach enables transferring the problem of analyzing complex real systems into a problem of solving a system of ordinary differential equations or a system of algebraic linear or non-linear equations. The latter occurs in case of modeling continuous-flow fermentation processes. Standard and relatively readily available software means can then be used for computer solution of a simulation problem. The use of structured, segregated models derived from population balances[57] is, because of highly demanding simulation software, rather an exceptional case in mathematical modeling of microbial processes.

Since problems and theoretical aspects of nonideal mixed system reactor modeling have been very thoroughly studied in connection with the development of chemical reactor engineering, a number of outstanding monographs exist covering the topic[34,41,88]. In the following section two examples are presented on the use of simple nonideal mixing models relevant to the type of problems often encountered in the area of microbial process engineering.

EXAMPLE 2.8

During fermentative production of citric acid by the fungal culture of *Aspergillus niger* a situation may occur whereby the culture starts growing in pellets. Devise a way of simulating differences and changes in the overall production rate as related to the pellet size.

Solution:

Biosynthesis of citric acid takes place usually under culture conditions when most of the substrate has been converted into the product and further growth is negligible with regard to the pellet size. The substrate and dissolved oxygen have to diffuse through the microbial biomass pellet where they are utilized for product formation. For description of the process a dispersion model without convection can be used. The diffusion rate under these conditions is characterized by effective diffusion coefficients D_S, D_P, D_O for the substrate, product and oxygen, respectively. Since the concentrations of substrates and product in the broth are changing with time only slowly, diffusion equations for individual time intervals of the fermentation can be solved for steady states. The model for non-ideal mixing within an individual pellet will assume the following form:

$$D_S \left[\frac{d^2 C_S}{dl^2} + \frac{2}{l} \frac{dC_S}{dl} \right] = r_S \, X$$

$$D_P \left[\frac{d^2 C_P}{dl^2} + \frac{2}{l} \frac{dC_P}{dl} \right] = -r_P \, X$$

$$D_O \left[\frac{d^2 C_O}{dl^2} + \frac{2}{l} \frac{dC_O}{dl} \right] = -Y_{O/P} \, r_P \, X$$

The boundary conditions for the centre of the pellet are determined from the pellet symmetry:

$$l = 0; \qquad \frac{dC_S}{dl} = \frac{dC_O}{dl} = \frac{dC_P}{dl} = 0$$

Second boundary condition for the surface of the pellet of diameter d is based on the concentrations of the product, substrate, and oxygen in the medium, respectively. It can be expressed as follows:

$$\text{for } l = \frac{d}{2} : \qquad C_S = C_{Sm} \qquad C_P = C_{Pm} \qquad C_O = C_{Om}$$

where index m denotes concentrations in the medium. Specific rates r_S and r_P depend on the concentrations of sugar and oxygen. The effectiveness of citric acid biosynthesis in dispersed or pellet culture form can now be characterized by an effectiveness factor defined as follows:

$$\eta = \frac{\int\limits_{0}^{d/2} 4\pi\, l^2\, r_P\, (C_S,\ C_O)\, X\, \mathrm{d}l}{(1/6)\, \pi\, d^3\, r_P\, (C_{Sm},\ C_{Om})\, X}$$

Models of a similar type are often used for description of microbial films and flocs[2]. They can be applied for analysis of surface cultivations or for reactors with immobilized cells.

EXAMPLE 2.9

Formulate a simple mathematical process model describing a continuous-flow culture for producing biomass on a viscous medium causing incomplete mixing in regions of the liquid broth volume in the bioreactor.

Solution:

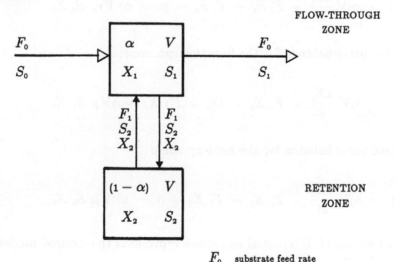

Figure 2.6:

Schematic diagram of an incompletely mixed reactor (after Sinclair and Brown[73]).

F_0	substrate feed rate
V	volume of the batch
α	flow-through volume of the batch
X	biomass concentration
S	substrate concentration
F_1	mass transfer rate between the flow-through and the retention zone.

A case of this sort was discussed in the paper by Sinclair and Brown[73]. The whole liquid broth volume was divided into two compartments with different substrate and biomass concentrations as seen in Figure 2.6. The species mass balances for a system characterized by specific growth rate m and specific substrate utilization rate r_s will assume the following form:

Substrate mass balance for the continuous flow-through zone:

$$\alpha V \frac{dS_1}{dt} = F_0 S_0 + F_1 S - (F_0 + F_1) S_1 - \alpha V r_s S_1 X_1$$

$(accumulation)$ = $(inflow)$ − $(outflow)$ − $(consumption)$

Substrate mass balance for the hold-up zone:

$$(1 - \alpha)V \frac{dS_2}{dt} = F_1 S_1 - F_1 S_2 - (1 - \alpha) V r_s S_2 X_2$$

Biomass mass balance for the flow-through zone:

$$\alpha V \frac{dX_1}{dt} = F_1 X_2 - (F_0 + F_1) X_1 + \alpha V \mu S_1 X_1$$

Biomass mass balance for the hold-up zone:

$$(1 - \alpha)V \frac{dX_2}{dt} = F_1 X_1 - F_1 X_2 + (1 - \alpha) V \mu S_2 X_2$$

The above set of differential equations represents the desired model of the given system.

The model of this type can be used as another alternative of a bioreactor with immobilized cells propagated on a growth-supporting medium, if the set of mass balances is supplemented by those concerning the product(s). Further illustrations and review of non-ideal mixing models for microbial processes can be found e.g. in the review paper by Ollis[52].

In the conclusion of this part devoted to the use of systems analysis for formulation of mathematical models of microbial processes it is necessary to emphasize that problems have been mainly discussed which are dealing with decomposition of a microbial process into individual subsystems. One of the reasons has been simplicity of the approach so as to suit the potential of available computer software means. Published literature usually contains only the complex and final versions of models used for process simulation purposes and the results of these simulations. However, the essential "know-how" of the model formulation and very often also the biological interpretation of individual model equations is either not described and discussed at all, or, if it is, only in a very superficial way. Very often there are a number of unsuccessful attempts before a suitable model is formulated, and these, although very educational and informative, will never appear in the published literature. The published final model and simulation results for the process of alkaloid biosynthesis by *Claviceps purpurea*[54], for example, was the eleventh alternative in a series of not so successful former attempts. Formulation of a suitable model for direct oxygen transfer in a microbial culture[75,76] was a result of painstaking processing of discussions and numerical analysis of experimental and production data over a period of three years.

A certain amount of attention in this work has been devoted to the discussion of significance and application of models of individual sub-processes in an attempt to create a suitable "dictionary" for translating the biological interpretation of experimental results into mathematical interpretation and vice versa. This was aimed at a reader with little experience in practical use of methods of mathematical modeling in engineering analysis of microbial processes. There is no suitable and instructive review text in the current literature devoted to the problems of "modeling know-how" as is the case in other engineering disciplines.

The model formulation problems dealt with in this part were discussed mainly from the viewpoint of process simulation. Process simulation, however, is the result of solving the entire model, i.e. model with all the constants, coefficients and/or parameters substituted by concrete numerical values. These values are derived by different ways of analyzing available experimental or operating data or reliable data reported in the accessible literature. Results of simulation computations should reflect behaviour of the real process as closely as possible.

Model identification is the global term used to designate the set of procedures and methods routinely applied in order to express the numerical values of model constants and parameters. It is quite possible that during this process some changes and modifications of the entire model itself may also be considered. When the model formulation is firmly established the task of model identification is reduced to the numerical expression of model parameter values most often through the use of "curve fitting" methods. In the following section a brief and practically oriented review of the "state of the art" in this area is presented.

3. MATHEMATICAL MODEL IDENTIFICATION

The formulation of mathematical fermentation process models, from the standpoint of systems analysis, is usually realized in three stages:

(a) qualitative analysis of the structure of a system, usually based on the knowledge of metabolic pathways and biogenesis of the desired product;

(b) formulation of the model in a general mathematical form. This stage is sometimes called the structure synthesis of the process functional operator[35];

(c) identification and determination of numerical values of model constants and/or parameters which is based on experimental or other operating data from a real process.

The previous sections of this text dealt with the first two points above concerning formulation of the mathematical model of a process. The ultimate goal of systems analysis is derivation of such a way of mathematical description of a process which would facilitate easier solution to process design problems, its optimal control, and overall optimization of production facilities. From long experience with processing of experimental data from fermentation processes it has been learned that when attempting formulation of a mathematical process model it is advisable to proceed in several sequential steps which eventually becomes almost routine.

First, numerical values of key process rates and stoichiometric coefficients are determined from experimental or process operating data. The *second* step is based on semiquantitative ways of evaluating the original information on the process mechanism from the experimental error point of view, and approximate numerical values of model parameters are estimated. In the *third* step of data processing a non-linear regression fit of the model on the original data is carried out.

3.1 PRELIMINARY ANALYSIS OF EXPERIMENTAL DATA

This phase of experimental data processing consists of verification of correctness and completeness of measured and recorded data by using the mass and energy balance method[14,48]. Expressing of the macroscopic process yield coefficients determined during fermentation experiments assists in the first-approach evaluation of the quality of processed experimental data and methods used for monitoring the process variables. It quite often happens that the thermodynamic efficiency of the process, the sum of η and ξ_P, when numerically expressed from the process data, results in values greater than 0.55 for aerobic and 0.7 for anaerobic processes. This is very often due to the omission of alternative carbon sources, substrates and media additives such as corn steep liquor, yeast extract or antifoam agents from the mass balance of the primary and easily measured culture substrates. These alternative potential carbon and energy sources can significantly affect the growth dynamics and biochemical production abilities of the microbial population. It is important to emphasize here that accurate determination of the yield coefficients is essential for successful formulation of the model in further steps of the procedure. As shown by the parametric sensitivity analysis of mathematical process models, the yield coefficients are invariably the most important model parameters. Stoichiometric coefficients express links and limitations of the whole process, as explained in Section 2.2, and therefore they could be used for the elimination of incomplete or erroneous data information in the model formulation procedure.

Apart from the evaluation of stoichiometric coefficients, evaluation of individual process rates characterizing the dynamic properties of individual subsystems in the process hierarchy is also part of the preliminary data analysis. When evaluating these rates a consideration has to be given to the conditions under which they were obtained. For the steady-state continuous-flow culture process, for instance, individual rates can be expressed by solving simple algebraic equations based on the mass balances for the bioreactor system. For time-variable systems the procedure is more involved and usually special algorithms for approximation by smooth curves need to be used for evaluation of derivatives resulting from the mass balance relationships. Table 3.1 presents a concise review of methods and computer programs for approximation of experimental data by smooth curves. Methods developed for this purpose are relatively complex and their application requires the use of a computer because smoothing is done on a relationship burdened by an experimental error ε_i. The situation can be expressed as

$$y_i = f_i(t) + \varepsilon_i \tag{3.1}$$

where y_i is the measured quantity and f is the real value which is represented during the smoothing operation by an approximating function. When using the algorithms for replacing experimental data points by a smooth curve, a dilemma is usually being resolved between the error of approximation, expressed by the sum of squares of deviations of experimental data from function $f_i(t)$, and a smooth relationship. The smoothness can be expressed mathematically as an integral over the changing square of curve derivatives and, consequently, approximation by a smooth curve can for instance be solved by using variable weighing coefficients p minimizing the relationship:

$$min \left\{ p \sum_{i=1}^{N} \left[\frac{y_i - f_i(t)}{\varepsilon_i} \right]^2 + (1 - p) \int_{t_1}^{t_N} [f^m(\tau)]^2 \, d\tau \right\} \tag{3.2}$$

This formulation of data approximation by a smooth curve was recommended by de Boor[5]. Table 3.1 also lists other references concerning this topic.

TABLE 3.1

**LIST OF PUBLISHED COMPUTER PROGRAMS FOR
EXPERIMENTAL DATA FITTING**

METHOD	PROGRAM TITLE	PROGRAMMING LANGUAGE	REFERENCE	AUTHOR
Gramm - first degree three point formula	SMOOTH 13 SMOOTH 35	ALGOL 60	Comm. ACM 6(1963),387.	Rodriguez-Gil, F.[63]
Fourth order smoothing	SMOOTH	ALGOL 60	Comm ACM 6(1963),663.	George, R.[20]
Cross spectrum smoothing via finite Fourier transformation	CSAD	FORTRAN IV	Appl. Stat. 23(1974),238.	Frome, E.L.[18]
Spline function	SMOOTH	ALGOL 60	Num. Math. 10(1967),182.	Reinsch, C.[59]
Cubic spline	DCSSMO DCSINT	FORTRAN IV	Trans ACM Soft. 6(1980),92.	Duris, C.[13]
Piecewise polynomial smooth curve fitting	SMOOTH	FORTRAN 77	*A practical guide to Splines*, Springer Verlag, N.Y. (1978).	de Boor, C.[5]
Cubic spline function	FLATC	FORTRAN IV	*Algorithmen zur Konstruktion Glatten Kurven und Flachen*, Oldenburgh Verlag, Munchen (1968).	Spath, H.[79]

EXAMPLE 3.1

During anaerobic microbial production of ethanol on a complex medium containing sugar the following experimental data were recorded from a batch culture process:

Time (h)	Concentrations (g/L) Biomass	Sugar	Ethanol
0	1.0	50	0.04
2	1.2	47	1
4	2.0	43	4
6	3.2	39	10
8	4.5	24	18
10	5.8	8	25
12	6.4	0	27
14	6.3	0	27

A preliminary data analysis should be done by using the method of mass and energy balances.

Solution:

For the determination of macroscopic yield coefficients the data points between time (0-12) h should be used because later on into the fermentation (hours 12-14) cell lysis invariably occurs. This aspect would be responsible for the distortion of process thermodynamic efficiency value considered with regard to biomass. From the process mass balance the yield coefficients can be calculated:

$$Y_{X/S} = \frac{6.3 - 1.0}{50 - 0} = 0.106 \qquad Y_{E/S} = \frac{27 - 0}{50 - 0} = 0.54$$

For calculation of the thermodynamic process efficiency with regard to biomass and product, Equations (2.11) and (2.12) can be used, respectively. Since the elemental composition of biomass in this experiment is not exactly known, the generalized biomass formula $CH_{2.1}O_{0.9}N_{0.15}$ suggested by Roels[64] can be adopted.

Parameters required for Equations (2.11) and (2.12) will be:

$$\sigma_X = 0.488 \qquad \sigma_S = 0.4 \qquad \sigma_P = 0.522$$
$$\gamma_X = 4.19 \qquad \gamma_S = 4 \qquad \gamma_P = 6$$

Details concerning the calculations of σ and γ can be seen in Example 2.4. The calculations of thermodynamic efficiencies η and ξ_P follow:

$$\eta = \frac{Y_{x/s}}{\dfrac{\sigma_s \, \gamma_s}{\sigma_x \, \gamma_x}} = \frac{0.106}{0.7825} = 0.135$$

$$\xi_P = \frac{Y_{P/s}}{\dfrac{\sigma_s \, \gamma_s}{\sigma_P \, \gamma_P}} = \frac{0.54}{0.51} = 1.058$$

Evaluation of the two efficiencies shows that $(\eta + \xi_P) \gg 0.7$ which is an indication of either incomplete or erroneous experimental data. Consequently, it is not advisable to consider formulation of a mathematical process model before the uncertainties concerning the macroscopic process mass balance are resolved. For the anaerobic fermentation of sugar into ethanol, Roels[64] presents the value of the overall thermodynamic process efficiency in the range from 0.62 to 0.68.

EXAMPLE 3.2

Upon evaluating the methodology of the experiment from preceding Example 3.1 a systematic error was revealed in the analytical determination of fermentable sugars in the broth. The actual concentration of fermentable sugars was 1.8 times higher. The overall thermodynamic process efficiency agrees with reported values[64] and it is 0.663.

Evaluate the numerical values of specific rates of growth and production based on the correct data.

Solution:

The numerical method for approximation of measured data by smooth curves published by de Boor[5] was applied in this case. Table 3.2 lists the smoothed values of experimental points as well as their first derivatives and specific rates of growth, substrate utilization, and product formation. From this Table also obvious is the transition from the lag to the exponential growth phase between hour 1 and 3 as well as the growth and product formation limitation by substrate occurring during the last three hours of the fermentation.

In general, for smoothing experimental relationships by numerical methods, it is possible to use results of the determination of process rates for semiquantitative evaluation of the mechanism because the error involved in the determination of a derivative can easily be up to 30% compared with the actual value.

TABLE 3.2

SOLUTION TO EXAMPLE 3.2
(COMPUTER PRINTOUT)

t[h]	X[g/L]	S[g/L]	P[g/L]
0.0	0.9247	90.0069	-0.0382
2.0	1.2811	83.9690	1.0161
4.0	2.0371	77.1078	4.0396
6.0	3.2029	69.8495	10.0408
8.0	4.5380	43.0276	18.0376
10.0	5.7086	14.0434	24.8250
12.0	6.3125	0.0304	27.0463
14.0	6.3951	-0.0347	27.0326

t	$\dfrac{dX}{dt}$	$\dfrac{dS}{dt}$	$\dfrac{dP}{dt}$
0.0	0.1376	-2.7482	0.3507
2.0	0.2594	-3.5605	0.8801
4.0	0.4936	-2.3586	2.2456
6.0	0.6491	-8.1845	3.6747
8.0	0.6613	-16.0238	4.0525
10.0	0.4641	-11.4295	2.2916
12.0	0.1439	-2.7540	0.2943
14.0	-0.0100	1.3282	-0.1575

t	$\dfrac{1}{X}\dfrac{dX}{dt}$	$-\dfrac{1}{X}\dfrac{dS}{dt}$	$\dfrac{1}{X}\dfrac{dP}{dt}$
0.0	0.1488	2.9719	0.3792
2.0	0.2025	2.7793	0.6870
4.0	0.2423	1.1578	1.1023
6.0	0.2027	2.5554	1.1473
8.0	0.1457	3.5310	0.8930
10.0	0.0813	2.0022	0.4014
12.0	0.0228	0.4363	0.0466
14.0	-0.0016	-0.2077	-0.0246

3.2   RATE RELATIONSHIPS AND KINETIC PARAMETERS

The choice of a suitable rate model was from a methodological point of view addressed in Section 2.1. Sometimes this aspect becomes relevant only when it comes to model identification. Either there is a fairly good idea as to what mechanism could well describe the dynamic behaviour and properties of the system in question or, and this is apparently a more frequent case, such an idea is to be derived from the data yielded from the preliminary analysis of experimental data. This is usually accomplished by plotting the derived numerical relationships for rates against concentrations of either the substrate or of the product. The plot and correlation of the rates against each other enables estimation of local yield coefficients or eventually of their mutual relationships. This approach assists in finding the most suitable relationships between the rates and concentrations.

Carrying out this step during the model identification procedure, whereby the possible type of a functional relationship is visually verified, serves as a very important stage in data processing during which a conclusion can be arrived at as to the possible mechanism of biosynthesis. Simultaneously, an order of magnitude estimation is carried out for the unknown kinetic model parameters which is subsequently used in the last phase of model identification.

EXAMPLE 3.3

Use the results from Example 3.2 for data analysis and for the suggestion of a possible mechanism describing the culture process dynamics. Carry out an approximate estimation of model parameters.

Solution:

As a first step, a graphical relationship is plotted of the specific growth rate μ and the substrate concentration S (Figure 3.1a). It is seen that the initial growth phase is delayed by the lag phase lasting for 1 to 2 hours. If these data are eliminated from consideration, the remaining relationship can be described by a function of r_1 type according to Table 2.1. The growth model can then be represented by a combination of two differential equations describing also the culture lag phase. Derivation of the model and more detailed discussion is presented in Section 2.3, Example 2.6. The resulting growth model will be

$$\frac{\mathrm{d}X}{\mathrm{d}t} = \left(\frac{\mathrm{RNA} - a}{b}\right) X$$

$$\frac{\mathrm{dRNA}}{\mathrm{d}t} = \left(k - \frac{1}{X}\frac{\mathrm{d}X}{\mathrm{d}t}\right) \mathrm{RNA}$$

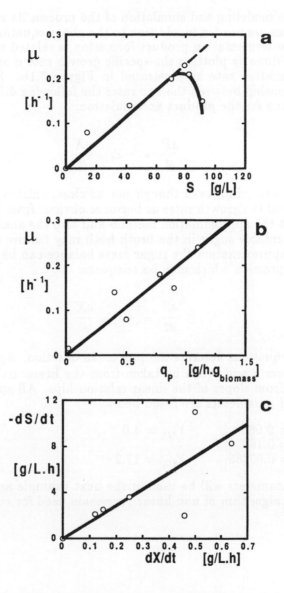

Figure 3.1(a-c): Selected relationships of fermentation process parameters for
constructing the mathematical model of the fermentation process in
Example 3.3 (results of relevant calculations are shown in Table 3.2):

a) Relationship of the specific growth rate (μ) and the substrate concentration (*S*).

b) Relationship of the growth rate (d*X*/d*t*) and the product formation rate (d*P*/d*t*).

c) Relationship of the growth rate (d*X*/d*t*) and the substrate utilization rate (d*S*/d*t*).

Since for the modeling and simulation of the process its sugar and product mass balances are needed in addition to the above equations, it is necessary to find if the fermentation product formation is related to culture growth. This can be done by plotting the specific growth rate μ against the specific product formation rate as illustrated in Figure 3.1b. From the obvious linear relationship between the two rates the following differential equation can be written for the product accumulation:

$$\frac{dP}{dt} = -Y_{P/X} \frac{dX}{dt}$$

A similar relationship, even though not as close, exists between the sugar utilization and the growth rates as becomes obvious from Figure 3.1c. Considering that the approximation method and also the analytical determination of fermentable sugar in the broth both may feature significant errors, in the first approximation the sugar mass balance can be simulated by the following expression which may be adequate:

$$\frac{dS}{dt} = -Y_{S/X} \frac{dX}{dt}$$

These four equations describe the process in question. Approximate values for parameters a and b can be taken from the literature[25], the others are determined from slopes of the linear relationships. All approximate values of the model parameters are listed below:

$a = 0.08$ $Y_{P/S} = 4.0$
$b = 0.16$
$k = 0.00285$ $Y_{S/X} = 17.2$

These parameters will be used in the next example as initial values for an iteration algorithm of non-linear regression used for curve fitting of the model.

3.3   EVALUATION OF MODEL PARAMETERS

Determination of numerical values of the mathematical model parameters depends on the adopted hierarchical level of structural relationships in the system and on the formal mathematical apparatus used for describing the model. Accordingly, an appropriate method of parameter determination is selected. Basically, the methods used can be divided into three groups:

(a) methods of linear and non-linear regression
(b) methods of the momentum analysis of experimental data
(c) methods of adaptive identification and estimation of model parameters.

Review papers by Himmelblau[30], Bard[4] and Kafarov[35] offer many details concerning ways of use and application of these methods with numerous examples. Since a more detailed elaboration of this topic would be beyond the scope of this text, the following paragraphs will be restricted only to a brief description of the above mentioned methods for determination of model parameters (coefficients).

a) Methods of non-linear and linear regression

are based on conventional methods of mathematical statistics and in practice they are the ones most often used. They are particularly useful for solution of problems concerning determination of parameters in models consisting of one or more linear or non-linear equations with constant parameters to be evaluated. In general, these methods are based on searching for a minimum of a certain criterion describing the agreement between experimental data and an approximation by the model. Most often different variants of the sum of weighed squares of deviations are used so that the criterion of agreement has the following form:

$$\left[\begin{array}{c} criterion\ of\ agreement \\ between\ model\ and\ data \end{array} \right] = \sum_{i=1}^{n} \sum_{j=1}^{m} w_{ij} \Delta_{ij}^2 \qquad (3.3)$$

where   n   ... number of unknowns in the model
    m   ... number of measured experimental data points
    w_{ij} ... weighing coefficient
    Δ_{ij} ... numerical value of the difference between the experimental point value and the value of dependent variable calculated from the model

Additional types of criteria are recommended in the literature such as Box determinant[4,30] or non-parametric criteria[30] but for practical calculations in most cases the use of the criterion based on the least squares of deviations (Equation 3.3) will prove adequate.

When processing data from fermentation processes, in most cases it is seen that the model is represented by a simple relationship explicitly expressing how the dependent and independent variables are related, or by a more complex formula whereby the model consists of a system of non-linear algebraic equations (continuous-flow cultures), or eventually of differential equations (batch and fed-batch cultures) with initial conditions. The model formulation with partial differential equations or with ordinary differential equations with boundary conditions is not very typical for problems concerning determination of kinetic parameters in models of fermentation processes and consequently it will not be dealt with in this text.

For the solution of simpler cases whereby only one relationship in the form of an explicit function is evaluated there are a number of computer programs and program systems described and readily available in the literature. Table 3.3 presents a brief review of some suitable computer programs from among many recently published which, however, may not have been verified by an independent researcher. The work of Hiebert[28] summarizes

TABLE 3.3

BRIEF REVIEW OF COMPUTER PROGRAMS FOR NON-LINEAR REGRESSION OF SIMPLE FUNCTIONS

METHOD	PROGRAM TITLE	PROGRAMMING LANGUAGE	REFERENCE	AUTHOR
Curve fitting with constraints	CURVE FITTING	ALGOL 60	Comm. ACM 5(1962),44.	Peck, J.E.L.[55]
Least squares solution with constraints	CONLSQ	ALGOL 60	Comm. ACM 6(1963),313.	Synge, M.J.[80]
Gram-Schmidt Transformation	GSFIT	FORTRAN IV	Com. Phys. Comm. 8(1974),56	O'Connell, M.J.[51]
Marquardt Method Powell Method	SSQMIN BOTM	FORTRAN IV	*Optimization Techniques in Fortran* McGraw Hill, N.Y. (1973)	Kuester, J.L., Mize, J.H.[39]
Nonlinear least squares estimation	TJMARI	FORTRAN IV	Sandia Tech. Rep. No SLL73-0303	Jeffreson, T.H.[33]

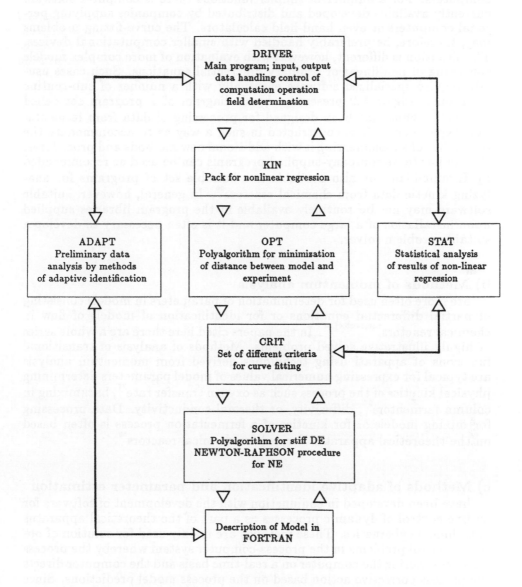

Figure 3.2: Block diagram of BIOKIN modular system of computer programs developed for estimation of parameters in the mathematical process model.

in quite an impartial way evaluation of existing programs for use on larger computers. For a number of simpler functions there is computer software currently available developed and distributed by companies supplying personal computers or even hand-held calculators. The curve-fitting problems may, therefore, be preferably handled with smaller computational devices. The situation is different, however, with evaluation of more complex models consisting of non-linear or ordinary differential equations. Such cases usually require specialized software eventually with a number of sub-routine programs. Figure 3.2 presents a block diagram of a program set called BIOKIN[85] which has been designed for processing of data from fermentation experiments. It is constructed in such a way as to accommodate the possibility of supplementing it with additional new methods and procedures. For similar tasks company-supplied programs can be used as recommended by Himmelblau[30] or also available is a suitable set of programs for analyzing kinetic data from chemical reactors[26]. In general, however, suitable software may not be routinely available in the program libraries supplied upon installation of a large computer and it is often necessary to develop a suitable problem solver.

b) Methods of momentum analysis

are more often used for determination of parameters in models consisting of partial differential equations or for identification of models of flow in chemical reactors[3,30,34,35,41,88]. In the papers cited here there are a whole series of highly illustrative solved problems. Methods of analysis of transitional functions of apparati using techniques derived from momentum analysis are typical for expressing numerical values of model parameters determining physical kinetics of the process such as oxygen transfer rate[77], backmixing in column fermentors[27], diffusivity, or thermal conductivity. Data processing for mixing models or for kinetics of a fermentation process is often based on the theoretical apparatus derived for chemical reactors[52].

c) Methods of adaptive identification and parameter estimation

have been developed in conjunction with the development of software for on-line control of dynamic processes as a part of the theoretical apparatus of technical cybernetics. These methods are mainly used for solution of optimal control problems in the process-computer system whereby the process data are entering the computer on a real-time basis and the computer directs the process corrective action based on the process model predictions. Since some of the system parameters, and therefore also the model parameters, can change in time, the computer continuously re-evaluates model parameters so that the behaviour of the process can be predicted and controlled for optimum performance. Methods of adaptive identification, however, can be used even in case of off-line experimental data analysis when some parameters may be changing in an unknown way. If their changes are to be evaluated, experimental data can be approximated by a suitable function (see Section 3.1) and then adaptive algorithms used for estimation of the variable value of the parameter. This method was used, for instance, for evaluation of aeration capacity of non-ideally mixed reactors[78]. The basic most frequently used methods such as the Kalman-Bucy filtration can be found in the above cited publications[30,34] while broader reviews have been presented by Rastrigin[58] or Iserman[32] and many others. The theory of adaptive identification is currently still developing with the expansion of microcomputers and its summary can hardly be finalized because it is also connected with the development of self-learning computing systems and robots. It is certain, however, that most of the methods successful in solution of cybernetic problems of identification are likely to be, in the near future, applicable also in the modeling of fermentation processes.

EXAMPLE 3.4

Data from Examples 3.2 and 3.3 are to be used for evaluation of process parameters described by the model formulated in Example 3.3.

Solution:

By using a computer program developed basically according to the block diagram presented in Figure 3.2 parameters were found corresponding to the minimum sum of weighed squares:

$$Criterion = \sum_{i=1}^{8} \left(S_{i\ calc} - S_{meas}\right)^2 \ 0.0001562 +$$

$$+ \sum_{i=1}^{8} \left(P_{i\ calc} - P_{meas}\right)^2 \ 0.0016 +$$

$$+ \sum_{i=1}^{8} \left(X_{1\ calc} - X_{meas}\right)^2 \ 0.0278$$

An iteration calculation, using a Rosenbrock method published by Kuester[39] for the criterion minimalization with regard to parameters, yielded the following values:

$$k = 0.0032 \qquad\qquad Y_{S/X} = 16.76$$
$$a = 0.08 \qquad\qquad\quad Y_{P/X} = 5.31$$
$$b = 0.0598 \qquad\qquad Criterion = 0.020399$$

For solving the system of mass balance equations a Runge-Kutta method of the 4th order was used which was published as algorithm no. 9 in Com. ACM **3**, 312 (1960). The solution of differential equations for the given parameters and initial conditions is presented in the following table:

Time [h]	Concentrations [g/L]			
	Biomass	Sugar	Ethanol	RNA
2.00	1.455	82.363	2.423	0.0953
4.00	2.383	66.818	7.355	0.0935
6.00	3.551	47.242	13.566	0.0901
8.00	4.688	28.175	19.616	0.0865
10.00	5.530	14.061	24.094	0.0835
12.00	6.012	5.993	26.654	0.0816
14.00	6.234	2.259	27.838	0.0806

Criterion: 0.020399

The results show that the model well coincides with the experimental data for the biomass and ethanol mass balances. For the sugar mass balance, agreement is not so good due to, as indicated earlier, the utilization of several sugars by the culture. In order to get a better agreement it would be necessary to supplement the model by mass balances of individual components.

Despite this slight inaccuracy, following its identification, the model can be used for simulation calculations for balancing the whole process line. The changes occurring in the values of parameters of this model version, as compared with preliminary evaluation in Example 3.2, are worth noticing. When starting the iteration calculations with values determined in Example 3.2 the criterion of agreement was higher by two orders of magnitude and the calculated concentration profiles for the product and substrate differed from the experimental values much more.

3.4   STATISTICAL EVALUATION OF MODEL IDENTIFICATION RESULTS

It is frequently necessary to evaluate the results of model identification from the standpoint of data analysis. Most of the routinely used methods are described in the review published by Himmelblau[30] and Bard[4]. A more detailed study of the series of Proceedings of Annual Winter Conferences on Simulation organized by the IEEE is also very stimulating as well as the comprehensive review by Shannon[71]. The most frequently discussed aspect is the question whether the results of data fitting are sufficiently good. Bard[4] recommends the solution of this problem based on using the "test for zero standard deviation" which is carried out in the following way:

The mean deviation of any dependent variable is determined from equation

$$\Delta_j = \frac{1}{n} \sum_{i=1}^{n} \Delta_{ij} \qquad (3.4)$$

The variation of residue error can then be according to this relationship

$$s_j = \frac{1}{h-1} \sum_{i=1}^{n} (\Delta_{ij} - \Delta_i)^2 \qquad (3.5)$$

For data characterization it is necessary to calculate λ which has an F-distribution:

$$\lambda = \frac{(n-m)n}{(n-1)m} \sum_{j=1}^{m} \frac{\Delta_j^2}{s_j} \qquad (3.6)$$

The truthfulness of the zero mean residue value hypothesis is then determined with the assistance of the F-distribution tables.

An example of the use of this technique can be seen, among others, in the study of microbial alkaloid formation[54]. This methodology became the one applied for data processing by the computer program set BIOKIN[85]. A computer program for testing of the regression model acceptability in FORTRAN IV language was published by Thomas[83] in the journal Applied Statistics. For the purpose of judging the model acceptability, this program is recommended in the collection of algorithms by the Communications of ACM.

Individual limits of confidence

of kinetic parameters represent another statistical characteristic describing the effect of individual kinetic parameters and their significance in the sequence of importance. The values can be determined numerically as was described in the solved problem of Himmelblau[30]:

$$\bar{k}_i = k_i \pm t_{1-\alpha/2}\, s\,(k_i) \tag{3.7}$$

where $\bar{k}_i$ is the possible value of the kinetic parameter for level of probability $(1 - \alpha)$, k_i is the value determined by non-linear regression, and $t_{1-\alpha/2}$ is a corresponding value of t-distribution. Quantity $s(k_i)$ is a parameter variance determined from a linearized covariant matrix.

Determination of individual limits of confidence enables analysis of the importance of individual fermentation model parameters and can be used in model reduction for application of an identified model in on-line adaptive control of the process by the computer. As a rule, the narrowest limits of confidence are associated with coefficients characterizing stoichiometric relationships among individual sub-systems, followed by kinetic constants and then eventually by parameters characterizing a limitation or inhibition.

In connection with formulation of mathematical models of fermentation processes the choice of parameters plays a very important role. It is generally thought that by adding one parameter into the model leads to an increased number of degrees of freedom and that almost anything can be described by using a higher number of parameters. This is a somewhat confusing statement. In the previously published work[76] concerning the direct oxygen transfer the model included a characteristic effect of flotation which resulted in a better correlation between the experimental data and the model.

The discrimination between two possible models

of the same process, so as to include the effect of an added model parameter on the number of degrees of freedom, can be done by applying the variance analysis in such a way that statistics λ is calculated from this relationship:

$$\lambda = \frac{VARIANCE_{model\ 2}}{VARIANCE_{model\ 1}} \geq F_{(1-\alpha)}(\gamma_1, \gamma_2) \tag{3.8}$$

where λ has F-distribution, and γ_1 and γ_2 are corresponding values of numbers of degree of freedom which are determined by the number of experimental points minus the number of parameters, $(1 - \alpha)$ is the probability of the hypothesis tested.

Applied statistics currently offers a number of other approaches. Their discussion would be well beyond the scope of this informative text and above cited monographs will serve as a good basis for further study with numerous relevant examples.

4. APPLICATION OF MATHEMATICAL MODELS IN THE SIMULATION AND OPTIMIZATION OF FERMENTATION PROCESSES

A mathematical model of the process is an extremely useful bioengineering tool. To develop this tool is not a self-serving procedure, the model should be used in a purposeful manner so that the time, energy, and capital expended on its development would bring an expected return and result in the desirable effect within the framework of the technological process for which the model was formulated. While learning aspects associated with pioneering a new development cannot be overlooked, there should always be a technological improvement resulting from application of the model. Furthermore, an indepth understanding of the methodology for process model development and its application is essential so that a lesson learned in one technological area where modeling has advanced could be translated into another one without undue "rediscovering of the wheel". This is particularly important with regard to fermentation processes and biological systems in general which have so far benefitted very little from modeling advances in the field of chemical reactor and process engineering, process modeling, simulation and optimization. The field is wide open for a significant contribution by students of open mind and interdisciplinary expertise.

4.1 SIMULATION

The simplest way of model application is its use in process simulation. In the simulation application, model equations are solved for different initial and boundary conditions according to a certain scenario based on the planning of simulation experiments. An undisputed advantage of a simulation experiment in fermentation processes is expedience with which a simulation can be carried out. While it takes hours or days to obtain results from a fermentation experiment, the answer based on a computer simulation can be gained almost instantaneously. Simulation studies enable testing of novel or unconventional technological variants of the process such as e.g. the change from a batch to a continuous-flow cultivation, or to the use of an immobilized-cell technology. For the bioreactor scale-up procedure, simulation, for instance, can easily test the process yield as affected by the mixing non-homogeneity in the large equipment. From the personnel training point of view, the process model can be used for simulation of new technological process conditions, abnormal or process failure conditions. Application of the process model in the latter case reveals its potential, so far not mentioned, which is in the "diagnostic" capacity of the model.

A mathematical model of the fermentation process can be described most frequently by one of the following ways as:

(a) a system of non-linear algebraic equations,
(b) a system of ordinary differential equations with initial conditions,
(c) a system of partial differential equations.

(a) Systems of non-linear algebraic equations

describing chemical or microbial processes in a steady state have to be solved by iteration methods. Table 3.3 presents a brief review of computer programs available for solution of problems described by a system of non-linear equations. A more detailed analysis of the key algorithms used for solution of these systems has been presented by Hiebert[29].

(b) Systems of ordinary differential equations (ODE) with initial conditions

are basically all dynamic models composed of ideally mixed compartments. These types of models are the most frequently encountered ones in the description of fermentation processes. Table 4.1 briefly lists the most important methods for solution of ordinary differential equations with initial conditions which have been published as independent computer programs.

TABLE 4.1

BRIEF REVIEW OF COMPUTER PROGRAMS FOR SOLUTION OF SETS OF ORDINARY DIFFERENTIAL EQUATIONS

METHOD	PROGRAM TITLE	PROGRAMMING LANGUAGE	REFERENCE	AUTHOR
Runge-Kutta 4th order	RK	ALGOL 60	Comm. ACM 3(1960),312.	Naur, P.[49]
Multistep methods of the predictor-corrector type	DIFSUB	FORTRAN IV	Comm. ACM 14(1971),185.	Gear, C.W.[19]
Runge-Kutta-Merson	MERSON	ALGOL 60	Num. Math. 14(1970),312.	Christiansen, J.[9]
Extrapolative	DIFFSYS	ALGOL 60	Num. Math 8(1966),1	Bulirsch, R.
Composite multistep	STINT	FORTRAN	ACM Trans. Math. Soft 4(1978),399.	Tendler, J.M. et al.[81]
Approximative derived from cubic splines	ADIP	FORTRAN	Comm. ACM 16(1973),635.	Burkowski, F.J.[7]

Most of the methods, however, have not been published in a computer program form and Table 4.1 thus represents only a fraction of approaches available for solution of ODE systems. There are currently 300-400 different methods described in available literature, but none of them could be labelled as a universal one particularly suitable for simulation of fermentation processes. This was the reason for a recent attempt[86] to develop a universal program system which would automatically select the most suitable method from a narrowed-down collection of eight most robust algorithms.

(c) Systems of partial differential equations

are typical for simulation problems with diffusional slowdown in bioreactors with immobilized cells or enzymes, for description of non-ideal mixing by a dispersion model and also for solution of problems with models of populations based on theory of population balances[57]. For solution of these problems there is specialized software as seen in Table 4.2 which, however, is not readily accessible for all potential users. Machura[43], Melgaard[46] and Dew[11] all presented more detailed literature reviews of this topic. Most of the software for solving partial differential equations, however, is already prepared on higher programming levels corresponding to simulation languages.

Simulation language is a higher form of a program

including an appropriate method of solving the model which effectively enables solution of the simulation problem regardless of the model type. Simulation languages have been developing simultaneously with programming languages from the 1960's.

Currently, there is a large number of simulation languages[65], however, only a few of them are readily available to the public because of their narrow specificity or dependency on a certain configuration of a computer. Further details concerning simulation languages can be found in specialized literature, e.g. good review work by Sargent[66] or a record from a panel discussion on the theme of "Teaching Simulation to Undergraduates" published in the Proceedings of the Winter Conference on Simulation[62] in 1982.

TABLE 4.2

**BRIEF REVIEW OF PUBLISHED PROGRAMMING SYSTEMS FOR SOLUTION
OF SETS OF PARTIAL DIFFERENTIAL EQUATIONS**

NAME	PROGRAMMING LANGUAGE	AVAILABILITY	AUTHOR AND REFERENCE
DSS/2	FORTRAN IV	For ~ $ 1,000	Schiesser, W.E.[67] Lehigh University Bethlehem, PA 18015, U.S.A.
PDEPACK	FORTRAN IV	no price quoted	Scientific Computing Consulting Services[69] 655 Allegheny Drive Colorado Springs, CO 80919, U.S.A.
PDCOL	FORTRAN IV	~ $ 50 from the ACM Algorithms Distributions Services	Madsen,N.K., Sincovec,R.F.[44] ACM Trans. Math. Soft. 5(1979),326.
PDETWO	FORTRAN IV	Same as PDCOL	Melgaard, D.K., Sincovec, R.F.[47] ACM Trans. Math. Soft. 7(1981),216.
DISPL	FORTRAN	National Energy Software Center, Argonne National Labs 9700 So. Cass Ave. Argonne, Ill. 60439	Leaf, G.[40] DISPL:A software package for one and two spatially dimensional kinetics diffusion problems. Report ANL-77-12
ITPACK 2A	FORTRAN	Dept. Num. Analysis, Univ. Texas, Austin, TX	Grimes, et al.[23] Report 164, Center for Num. Anal. Univ. Texas, Austin, TX

EXAMPLE 4.1

Carry out simulation calculations with the bioreactor model described in Example 2.6.

Solution:

This is a simple process model whereby an integration program, e.g. the one by Naur[49] modified as listed in Table 7.4 in the Appendix to Part III, can be used. The program was transcribed from language Algol 60 into FORTRAN IV and the main program RK was renamed ODE. Integration with accuracy ϵ over the interval t_1, t_2 will be done in such a way that on the input the initial conditions are deposited in vector y and on the output there is a solution which is then printed out.

The model is described by a sub-program for numerical expression of right-hand sides of the system of differential equations (RHS):

```
DIMENSION Y(20)
COMMON/PAR/AMI
READ (5,*) Y(1), Y(20)
TF = 0.0
AMI = 0.0
DO 1 I = 1,51
TB = TF
TF = TF + 0.25
WRITE (6,100) TF, Y(1), Y(2), AMI
FORMAT (1X, F5, 2, 2X, 3F 10.5)
CALL ODE (2, TB, TF, Y, 0.005)
STOP
END
SUBROUTINE RHS (N, X, Y, F)
DIMENSION Y(20), F(20)
COMMON/PAR/AMI
AMI = (Y(2) - 0.08)/0.16
F(1) = AMI * Y(1)
F(2) = (0.5-AMI) * Y(2)
RETURN
END
```

The simulation results are illustrated in Table 4.3.

TABLE 4.3

SIMULATION RESULTS FOR EXAMPLE 4.1

TIME (h)	BIOMASS (g/L)	RNA	μ (1/h)
0.25	0.100	0.080	0.000
0.50	0.100	0.089	0.062
0.75	0.103	0.099	0.122
1.00	0.107	0.108	0.179
1.25	0.112	0.116	0.231
1.50	0.120	0.124	0.277
1.75	0.129	0.130	0.317
2.00	0.140	0.136	0.351
2.25	0.154	0.140	0.380
2.50	0.170	0.144	0.404
2.75	0.188	0.147	0.424
3.00	0.210	0.150	0.439
3.25	0.235	0.152	0.452
3.50	0.263	0.154	0.462
3.75	0.296	0.155	0.470
4.00	0.333	0.156	0.476
4.25	0.376	0.157	0.481
4.50	0.424	0.157	0.485
4.75	0.479	0.158	0.488
5.00	0.542	0.158	0.491
5.25	0.613	0.158	0.493
5.50	0.693	0.159	0.494
5.75	0.785	0.159	0.495
6.00	0.888	0.159	0.496
6.25	1.006	0.159	0.497
6.50	1.139	0.159	0.498
6.75	1.291	0.159	0.498
7.00	1.462	0.159	0.498
7.25	1.656	0.159	0.499
7.50	1.876	0.159	0.499
7.75	2.126	0.159	0.499
8.00	2.409	0.159	0.499
8.25	2.729	0.159	0.499
8.50	3.092	0.159	0.499
8.75	3.504	0.159	0.499
9.00	3.970	0.159	0.499
9.25	4.499	0.159	0.499
9.50	5.098	0.159	0.499
9.75	5.776	0.159	0.499
10.00	6.545	0.159	0.499

The integration program used here was in the simulation program called by instruction CALL ODE (n, t_1, t_2, y, E) where n is the number of equations (n=2). The simulation results are depicted in Figure 2.5.

4.2 PARAMETRIC SENSITIVITY OF A FERMENTATION PROCESS

When mathematical process models are being used for process simulation studies aimed at improving current technology, the effect of individual parameters on the process can be assessed by using the parameter effect "mapping". Considering that complex processes usually have a whole series of elective parameters it may be advantageous to first estimate the degree of importance of individual parameters. For expedient assessment of the degree of importance of individual parameters in a technological process the method of parametric sensitivity has been widely used particularly in the chemical industry. This method is considered as one of the mathematical means for semi-quantitative process optimization because, for a given technology, it can specify the sequence of importance and the potential impact of individual parameters on the result of the entire process. An identified mathematical model can be used for calculation of the absolute parametric sensitivity of the process it describes by using the following relationship:

$$Absolute\ Parametric\ Sensitivity\ =\ \frac{\partial f}{\partial P}\ \approx\ \frac{\Delta f}{\Delta P}$$

where f designates the optimized function such as e.g. the concentration or productivity of antibiotics or other economically important process parameter. The initial concentration of one of the key medium components, or the temperature, feed rate, etc., can be taken as technological parameters. In order to follow individual parametric sensitivities it is recommended to use the relative parametric sensitivity defined as

$$Relative\ Parametric\ Sensitivity\ =\ \left|\frac{P\partial P}{f\partial P}\right|\ \approx\ \left|\frac{P\Delta f}{f\Delta P}\right|$$

When evaluating the results obtained from expressing the parametric sensitivity, the parameters are first organized according to values of their respective relative parametric sensitivities. This established the sequence of their importance in the technological process. Eventually, the value of the absolute parametric sensitivity is taken into consideration. Its sign characterizes the direction of the effect exhibited by the technological parameter. A positive sign indicates process improvement with the increasing value of the parameter. The opposite is the case with a negative sign.

Testing and a quantitative expression of the effect of technological parameters by applying the method of parametric sensitivity does not require excessive computer time. This operation should be carried out before computerized synthesis of the optimal process configuration because it enables selection of the most important process parameters whereby the less important ones can remain constant for optimalization calculations with resulting decrease in the amount of calculations and saving of computer time. Following the process optimization it is advisable to check for eventual changes in the process parametric sensitivity.

EXAMPLE 4.2

A fed-batch culture process is to be simulated on a computer using culture kinetics from Examples 3.2 and 3.3. The feed is to be based on five (5) additions of 10% sugar solution equalling 0.1 of the initial batch volume each. The feeding is to be carried out when the sugar concentration in the broth decreases to 2 g/L on the whole hour of the fermentation time. The fermentation is considered finished when the broth sugar concentration after the five additions reaches the 2 g/L.

The initial biomass concentration (after inoculation) is 1 g/L, the starting sugar concentration is 25 g/L. Determine the process parametric sensitivity with regard to the number of feedings, feed sugar concentrations, and the initial sugar biomass and concentrations.

Solution:

The amount of product divided by the overall fermentation time can be selected as a suitable evaluation function reflecting the process productivity. In the first stage of solution the process simulation model will be formulated. The simulation program and results of the simulation are presented in the following paragraph. As process technological parameters the following quantities have been selected:

P_1	initial biomass concentration	1 g/L
P_2	initial sugar concentration in the broth	25 g/L
P_3	number of feed additions	5
P_4	the feed addition volume as a fraction of total volume	0.1
P_5	sugar concentration in the feed stream	100 g/L

The program used for simulation of the fed-batch culture:

```
          DIMENSION P(5), Y(20)
          READ(5,*) (p(i),i=1,5)
          TF=0.0
          Y(1)=P(1)
          Y(2)=P(2)
          Y(3)=0.0
          Y(4)=0.08
          FED=0.0
          DO 1 I=1,100
          TB=TF
          TF=TF+1.
          WRITE(6,100) TB,(Y(L),L=1,4)
   100    FORMAT(1Z,F5.2,1X,4F8,4)
          IF(FED.LT.P(3)) GO TO 2
          IF(Y(2).LE.2.1) GO TO 15
   2      IF(Y(2).GT.2.) GO TO 1
          FED=FED+1
          V1=FED * P(4)+1.
          V2=V1-P(4)
          Y(1)=Y(1) * V2/V1
          Y(2)=Y(2) * V2/V1+P(4)* P(5)/V1
```

```
          Y(3)=Y(3) * V2/V1
          WRITE(6,101)
101       FORMAT(' ****  FEEDING  ****')
          WRITE(6,100) TB,(Y(L),L=1,4)
1         CALL ODE(4,TB,TF,Y,0.01)
15        F=Y(2) * (FED * P(4)+1.)/TB
          RETURN
          END
          SUBROUTINE RHS(N,X,Y,F)
          DIMENSION Y(20),F(20)
          F(1)=(Y(4)-0.08) * y(1)/0.0598
          F(2)=-16.76 * F(1)
          F(3)=5.31 * F(1)
          F(4)=(0.0032 * Y(2)-F(1)/Y(1)) * Y(4)
          RETURN
          END
```

The main program carries out the simulation calculation assisted by the integration program ODE. The model is described by the sub-program RHS.

The program used for calculation of process parametric sensitivity:

```
          DIMENSION P(5),DP(5)
          READ(5,*) (P(I),I=1,5)
          READ(5,*) (DP(I),I=1,5)
          CALL PROC(P,FO)
          DO 1 I=1,5
          P(I)=P(I)+DP(I)
          ASENS=(F-FO)/DP(I)
          RSENS=ABS(ASENS * P(I)/FO)
1         WRITE(6,100) P(I),ASENS,RSENS
100       FORMAT(1X,3F10.5)
          STOP
          END
          SUBROUTINE PROC(P.F)
          DIMENSION Y(20),P(5)
          TF=0.0
          Y(1)=P(1)
          Y(2)=P(2)
          Y(3)=0.0
          Y(4)=0.08
          FED=0.0
          DO 1 I=1,50
          TB=TF
          IF(FED.LT.P(3) GO TO 2
          IF(Y(2).LT.2.1) GO TO 15
2         IF(Y(2).GT.2) GO TO 1
          FED=FED+1.
          V1=FED * P(4)+1
          V2=V1-P(4)
          Y(1)=Y(1) * V2/V1
          Y(2)=Y(2) * V2/V1+P(4) * P(5)/V1
          Y(3)=Y(3) * V2/V1
1         CALL ODE(4,TB,TF,Y,1.01)
```

```
15    F=Y(2) * (FED * P(4)+1.)/TB
      RETURN
      END
      SUBROUTINE RHS(N,X,Y,F)
      DIMENSION Y(20),F(20)
      F(1)=(Y(4)-0.08) * Y(1)/0.0598
      F(2)=-16.76 * F(1)
      F(3)=5.31 * F(1)
      F(4)=(0.0032 * Y(2)-F(1)/Y(1)) * Y(4)
      RETURN
      END
```

The parametric sensitivity program was derived as modification of the process simulation program. The whole printout of the program is listed here so that it could be seen how a simulation program can be modified for evaluation of the process parametric sensitivity.

The following Table presents the results of calculations expressing the parametric sensitivity for the given process.

Index	Parameter P	$\dfrac{\Delta f}{\Delta P}$	$\left\lvert \dfrac{P}{f} \dfrac{\Delta f}{\Delta P} \right\rvert$
1	1.00000	-0.81251	4.13424
2	25.00000	-0.01778	2.26129
3	5.00000	0.00000	0.00001
4	0.10000	5.08799	2.58889
5	100.00000	0.00783	3.98526

From the results it is seen that the most important ones in the hierarchy of technological parameters, according to the relative parametric sensitivity, are the initial biomass concentration and the sugar feed concentration. The feed addition size and the initial sugar concentration do have some significance although it is somewhat lower. Relatively unimportant, for the given technology, is the number of feed additions. Lowering of the initial biomass and sugar concentrations could lead to a higher productivity of the process, similarly so could increasing the feed rate and the feed sugar concentration.

TABLE 4.4

FED-BATCH PROCESS MODEL SIMULATION RESULTS

TIME (h)	BIOMASS (g/L)	SUGAR (g/L)	PRODUCT (g/L)	RNA
0.00	1.0000	25.0000	0.0000	0.0800
1.00	1.0367	24.3838	0.1951	0.0835
2.00	1.1088	23.1760	0.5777	0.0843
3.00	1.1916	21.7875	1.0175	0.0843
4.00	1.2778	20.3418	1.4755	0.0841
5.00	1.3648	18.8836	1.9373	0.0838
6.00	1.4511	17.4359	2.3959	0.0835
7.00	1.5359	16.0148	2.8461	0.0833
8.00	1.6182	14.6342	3.2834	0.0830
9.00	1.6975	13.3058	3.7043	0.0827
10.00	1.7730	12.0395	4.1055	0.0825
11.00	1.8444	10.8432	4.4845	0.0822
12.00	1.9112	9.7228	4.8394	0.0820
13.00	1.9733	8.6831	5.1691	0.0818
14.00	2.0305	7.7228	5.4730	0.0816
15.00	2.0828	6.8448	5.7511	0.0814
16.00	2.1304	6.0467	6.0040	0.0813
17.00	2.1734	5.3256	6.2324	0.0811
18.00	2.2121	4.6776	6.4377	0.0810
19.00	2.2466	4.0984	6.6212	0.0809
20.00	2.2774	3.5830	6.7844	0.0808
21.00	2.3046	3.1263	6.9291	0.0807
22.00	2.3286	2.7231	7.0568	0.0800
23.00	2.3498	2.3683	7.1692	0.0805
24.00	2.3683	2.0569	7.2679	0.0804
25.00	2.3846	1.7843	7.3542	0.0804
		FEEDING		
25.00	2.1678	10.7130	6.6856	0.0804
26.00	2.2087	10.0271	6.9029	0.0816
27.00	2.2722	8.9633	7.2399	0.0817
28.00	2.3373	7.8705	7.5861	0.0816
29.00	2.3982	6.8503	7.9093	0.0814
30.00	2.4531	5.9296	8.2010	0.0813
31.00	2.5019	5.1114	8.4602	0.0811
32.00	2.5448	4.3913	8.6883	0.0809
33.00	2.5824	3.7618	8.8878	0.0808
34.00	2.6150	3.2144	9.0612	0.0807
35.00	2.6433	2.7405	9.2113	0.0806
36.00	2.6676	2.3321	9.3406	0.0805
37.00	2.6885	1.9814	9.4517	0.0804

TIME (h)	BIOMASS (g/L)	SUGAR (g/L)	PRODUCT (g/L)	RNA
		FEEDING		
37.00	2.4645	10.1496	8.6641	0.0804
38.00	2.5100	9.3867	8.9058	0.0815
39.00	2.5778	8.2490	9.2662	0.0816
40.00	2.6461	7.1040	9.6290	0.0815
41.00	2.7086	6.0570	9.9606	0.0813
42.00	2.7637	5.1327	10.2535	0.0811
43.00	2.8116	4.3301	10.5077	0.0809
44.00	2.8527	3.6402	10.7263	0.0808
45.00	2.8879	3.0512	10.9129	0.0807
46.00	2.9177	2.5512	11.0712	0.0806
47.00	2.9429	2.1288	11.2050	0.0805
48.00	2.9640	1.7732	11.3177	0.0804
		FEEDING		
48.00	2.7360	9.3292	10.4471	0.0804
49.00	2.7822	8.5545	10.6925	0.0814
50.00	2.8506	7.4077	11.0558	0.0815
51.00	2.9185	6.2694	11.4164	0.0813
52.00	2.9795	5.2467	11.7404	0.0811
53.00	3.0323	4.3617	12.0208	0.0810
54.00	3.0772	3.6084	12.2591	0.0808
55.00	3.1149	2.9763	12.4597	0.0807
56.00	3.1465	2.4473	12.6272	0.0805
57.00	3.1727	2.0078	12.7664	0.0804
58.00	3.1944	1.6442	12.8816	0.0804
		FEEDING		
58.00	2.9662	8.6696	11.9615	0.0804
59.00	3.0127	7.8891	12.2087	0.0813
60.00	3.0810	6.7448	12.5713	0.0813
61.00	3.1479	5.6230	12.9267	0.0812
62.00	3.2071	4.6305	13.2411	0.0810
63.00	3.2575	3.7861	13.5086	0.0808
64.00	3.2995	3.0810	13.7320	0.0807
65.00	3.3343	2.4983	13.9166	0.0806
66.00	3.3628	2.0203	14.0680	0.0805
67.00	3.3860	1.6302	14.1915	0.0804
		FEEDING		
67.00	3.1603	8.1882	13.2454	0.0804
68.00	3.2077	7.3938	13.4971	0.0812
69.00	3.2759	6.2498	13.8595	0.0813
70.00	3.3420	5.1421	14.2104	0.0811
71.00	3.3996	4.1754	14.5167	0.0809
72.00	3.4480	3.3650	14.7734	0.0808
73.00	3.4877	2.6985	14.9846	0.0806
74.00	3.5200	2.1563	15.1563	0.0805

4.3   PROCESS OPTIMIZATION BY USING THE MODEL

Through the use of process simulation techniques, by manipulating the process technological parameters, it is possible to find an optimal or pseudo-optimal operating regime. The heuristical method of trial and error which is used in such an approach is basically one of the oldest optimization methods. The person or operator using the simulation program in this manner acts according to a certain, usually relatively simple, algorithm. If the person is replaced by an automatic optimization program, the situation arises whereby the optimal process synthesis is carried out by using a computer. A draft of a block diagram for such a program system is illustrated in Figure 4.1. The first block generates or reads the input parameters of the process. The second block realizes the selection of optimal process parameters either based on an analytical break-down of simple cases or by numerical methods. Optimization theory and applications have currently been developing at an accelerating pace and this section will present only a brief discussion of the main trends and techniques in optimization of technological processes by using process models. This will be done specifically with respect to the use of these techniques in the optimization of fermentation processes.

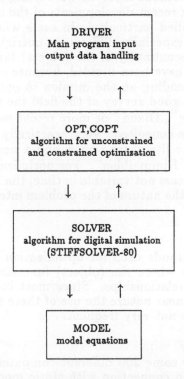

Figure 4.1: Modular system of computer programs for synthesis of the optimal technological conditions for a fermentation process.

Analytical methods of optimization

are based on the mathematical theory of extremes of smooth continuous functions. The optimum given by an extreme of the object function is obtained upon derivatization of this function by the technological parameter and the point is considered where the corresponding derivative is equal to zero. The resulting system of equations is subsequently solved and the solution represents the extreme of the (object) function. From the sign of the second derivative (in the location of the extreme) it can be decided if there is a maximum, minimum or an inflection point. If there are some limitations of the function or parameters, given for instance by the process stoichiometry, the function extreme can be located by applying the method of Lagrange multipliers. When solving real problems encountered in the optimization of technological parameters, particularly if the optimization result is supposed to be a certain value as opposed to an analytical relationship of the optimum with a technological parameter, a more rational approach is then based on a multipurpose method of numerical optimization.

Numerical optimization

and corresponding methods involved were being established simultaneously with the relatively recent developments of the systems theory. These methods have been applied particularly in cases where the analytical optimization approach is especially difficult or outright impossible because of the complexity or discontinuity of the (object) function. Techniques of numerical optimization have been divided, because of rapid specialization, into several groups depending on the mission to optimize either static or dynamic systems. For a good survey of the field the review publications of Gottfried and Weismann[22], Dixon[12] or more recent works of Bradley *et al.*[6] or Gill *et al.*[21] should be consulted. Very practically directed publications with an almost complete literature review and a number of examples are those by Schwefel[68] and Himmelblau[31]. For optimization of static models, i.e. models with parameters not variable in time, the known techniques can be divided according to the nature of the problem into linear and non-linear ones.

Linear programming

is a collection of methods used for optimization of complex economic and transport problems where the (object) function and constraints can be described by linear relationships. Since most models of fermentation processes are of a non-linear nature the use of these techniques for process optimization purposes is not very frequent.

Non-linear programming

is a general label for some 200 different computer methods for solving optimization problems in connection with static models. Most of the fermentation process optimization problems can be solved by one or more of these methods.   Table 4.5 summarizes some certified methods and algorithms of these methods suitable for a computerized approach to the solution. For comparison and testing of efficiency of individual techniques the publications of Schwefel[68] and Himmelblau[31] should be consulted, presenting some of the more important methods in the form of computer programs in FORTRAN IV. Some more optimization programs can also be found in the book by Kuester[39]. These programs, however, have not been tested.

Most of the above mentioned programs are meant for use on mainframe computers. Conley[10] shows in his book a number of examples from different fields on the use of optimization techniques for microcomputers equipped by a translator from BASIC.

TABLE 4.5

BRIEF REVIEW OF TESTED ALGORITHMS FOR OPTIMIZATION

PROGRAM TITLE	PROGRAMMING LANGUAGE	REFERENCE	AUTHOR
MINIFUN	ALGOL 60	Comm. ACM 5(1962),550.	Whitley, V.W.[89]
DIRECT SEARCH	ALGOL 60	Comm. ACM 6(1963),313.	Kaupe, A.F.[38]
STEEP 1 STEEP 2	ALGOL 60	Comm. ACM 6(1963),517.	Wasscher, E.J.[87]
DAPODMIN	ALGOL 60	Comp. J. 13(1970),111.	Sill, A.L.[72]
NELMIN	FORTRAN IV	Appl. Stat. 20(1971),338.	O'Neill, R.[53]
ROMIN	FORTRAN IV	Comm. ACM 16(1973),482.	Machura, M.[42]
JCONSX	FORTRAN IV	Comm. ACM 16(1973),487.	Richardson, J.A.[61]
MINI	FORTRAN IV	ACM Trans. Math. Soft. 2(1076),87.	Shanno, D.F.[70]

Dynamic programming

is an optimization technique which decomposes complex problems into simpler ones and it is typical for the solution of optimal performance of multistage systems. It is often used for solving transportation problems to find the shortest routes, or for optimization of large technological production units interconnected by complex relationships such as those based on mass and energy recycling. In chemical engineering practices this technique is also used for optimization of reactors and multistage separation processes.

In the area of fermentation process optimization it seems that this technique has a potential for optimization of the whole process including the fermentation stage. A general program for the solution of dynamic programming problems can be found, for instance, in the book by Kuester[39]. The method of dynamic programming introduced by Richard Bellman is sometimes also perceived as a discrete analogy of optimization methods for dynamic systems derived from the Pontyagin's principle of maximum, as presented by Kamien[36] and Fan[15].

Optimization using Pontyagin's principle of maximum

serves to locate the optimum performance of systems with dynamic parameters. The relevant theoretical basis can be found for instance in the book by Fan[15]. This method was originally proposed as a technique for optimal process control by a computer. It can be applied, with certain modifications, in off-line process optimization using a model, particularly for determination of the optimal temperature or pH profile, or the substrate feeding schedule. A certain disadvantage of the optimization methods derived from the principle of maximum is in a relatively tedious preparation of the model for solution of the boundary problem on a computer because for such problems there are very few tested and reliable software packages accessible in the literature as compared to the previously discussed methods. In the field of optimal control design, the methods derived from the principle of maximum are being displaced by heuristical self-learning algorithms of optimal control mentioned earlier in this text[32].

During optimization of complex technological entities a situation can arise where simultaneous optimization is required of several (value) functions which cannot be incorporated in one. This situation is described as the *vector optimization* or, not so accurately, as the *game theory*[37]. The formulation of corresponding algorithms, as has been done in previous cases, depends on the problem being of a static or dynamic nature, and also on the relationships among individual (object) functions which can be antagonistic, or competitive, etc. It seems that the current trends point to the potential future use of these methods in explanation and modeling of self-organizing systems. In the field of biology these methods may find a use in addressing problems of selection, competition of species in ecosystems, changes in the metabolic activities of microbial populations caused by environmental changes, and eventually also in problems associated with modeling of the process of evolution[60].

EXAMPLE 4.3

Determine the optimum profile for the feed rate of a sugar-based culture medium used in a 40-hour fed-batch anaerobic process of neutral solvents biosynthesis by *Clostridium acetobutylicum* bacteria. It is also desirable to maximize the yield of the neutral solvents produced.

Solution:

This problem is based on a real process of butanol biosynthesis whereby sugar is converted into biomass, butanol, acetone, and ethanol through intermediate metabolites, acetic and butyric acids, which are also found in the broth. From experimental batch data a mathematical model of the process was derived and identified (see also PART III of this text, the A-B-E Case Study). The following kinetic equations have been devised for individual metabolic sub-systems:

Specific rate of sugar consumption:

$$r_S = k_3 S + k_4 \frac{S}{S + K_{S1}}$$

Specific rate of butyric acid (P_1) production:

$$r_{BA} = k_5 S \frac{K_I}{K_I + P_2} - k_6 \frac{P_1}{P_1 + K_{S2}}$$

Specific rate of butanol (P_2) production:

$$r_B = k_7 S - 0.841\ r_{BA}$$

Specific rate of acetic acid (P_3) production:

$$r_{AA} = k_8 \frac{S}{S + K_{S1}} \frac{K_I}{K_I + P_2} - k_9 \frac{P_3}{P_3 + K_{S3}}$$

Specific rate of acetone (P_4) production:

$$r_A = k_{10} \frac{S}{S + K_{S1}} - 0.5 \, r_{AA}$$

Specific rate of ethanol (P_5) production:

$$r_E = k_{11} \frac{S}{S + K_S}$$

For expressing the specific growth rate, changes in the culture dynamics during the lag phase should be considered. In order to incorporate them in the model a similar relationship as in Example 2.6 will be used:

$$\mu = (RNA - a)/b - k_2 P_2$$

$$r_{RNA} = (k_1 S \frac{K_I}{K_I + P_2} - \frac{RNA - a}{b}) \, RNA$$

By using the procedure described in the previous sections of this text and based on the analysis of original experimental kinetic data the values of rate constants have been determined:

$k_1 = 0.0076$	$k_7 = 0.0107$	$K_{S1} = 2.0$
$k_2 = 0.0014$	$k_8 = 0.22$	$K_{S2} = 0$
$k_3 = 0.024$	$k_9 = 0.01$	$K_{S3} = 0.5$
$k_4 = 0.6$	$k_{10} = 0.132$	$k_I = 0.083$
$k_5 = 0.0135$	$k_{11} = 0.026$	$a = 0.08$
$k_6 = 0.1$		$b = 0.16$

The mathematical model of the fed-batch cultivation was derived from the differential mass balances:

$$\frac{dX}{dt} = \mu X - \frac{\dot{F}}{V} X$$

$$\frac{dS}{dt} = -r_S X - \frac{\dot{F}}{V}(S_0 - S)$$

$$\frac{dP_1}{dt} = r_{BA} X - \frac{\dot{F}}{V} P_1$$

$$\frac{dP_2}{dt} = r_B X - \frac{\dot{F}}{V} P_2$$

$$\frac{dP_3}{dt} = r_{AA} X - \frac{\dot{F}}{V} P_3$$

$$\frac{dP_4}{dt} = r_A X - \frac{\dot{F}}{V} P_4$$

$$\frac{dP_5}{dt} = r_E X - \frac{\dot{F}}{V} P_5$$

$$\frac{dV}{dt} = F(feed\ flow\ rate)$$

The carbon mass balance of the bioreactor system was used for formulation of a function to be optimized. This function can be expressed for example as a carbon balance-based sum of the desired products accumulation $(+)$ with the undesirable ones, including the residual substrate, subtracted $(-)$. The final broth volume when the fermentation terminated is considerable. The "optimized function" chosen for optimization of the acetone-butanol-ethanol biosynthesis process has the following format:

$$f_{OPT} = (0.6486 P_2 + 0.622 P_4 + 0.5217 P_5 - 0.4806 X - 0.4 P_3 - 0.4 S - 0.4865 P_1)\frac{V_i}{V_f}$$

where V_i/V_f is the ratio of the initial over the final volume of fermentation broth.

Since the problem of finding the optimal feed-rate profile represents a problem with dynamically changing parameters, a choice can be made as to the solution method applied. It can be based on the method of dynamic programming derived from Pontyagin's principle of maximum. Or, the problem can be transformed into an optimization problem with static parameters.

Considering that the model is represented by a system of non-linear differential equations it would very cumbersome to prepare it in a suitable form for application of dynamic methods of optimization. Instead, the dynamic problem of the optimal feed rate profile synthesis presented here should be transformed into a static one by replacing the unknown function of time with a series for the point where the formulation terminates and thus $F = 0$:

$$F = F(t_F) + \frac{\mathrm{d}F(t_F)}{\mathrm{d}t}(t - t_F) + \frac{1}{2!}\frac{\mathrm{d}F^2(t)}{\mathrm{d}t}(t - t_F)^2 + \cdots$$

From the optimization point of view a vector of static parameters a_1, a_2, a_3, $\cdots$ has to be determined with regard to the maximum of function f_{OPT} and for the constraint condition $F \geq 0$. The optimal feed rate profile can then be approximated by a function

$$F = a_1(t - t_F) + a_2(t - t_F)^2 + a_3(t - t_F)^3 + \cdots$$

Optimization of the feed rate profile was done separately and in a sequence for one, two, and three parameters a, respectively. The results for $t_F = 40$ and $S_o = 400$ g/L are presented in the following Table:

Number of parameters	a_1	a_2	a_3	f_{OPT}
1	-3.359×10^{-4}	–	–	26.42
2	1.781×10^{-3}	7.352×10^{-5}	–	35.54
3	$1.79 \ \times 10^{-3}$	7.156×10^{-5}	-5.265×10^{-8}	35.57

From the results of the synthesis of the optimal feed flow rate profile based on the use of the process model it can be seen that the optimal feed rate profile can be described by a parabolic function. The synthesis of the optimal feed flow rate profile is adequately done by a function describing the time variability of the feed which consists of just two or three members. It may be useful to also refer to the discussion of this case presented in PART III (The A-B-E Case Study), Section 7.3 (The Fed-Batch Culture System). Figure 7.34 shows a graphical illustration of simulation results for a fed-batch cultivation with the optimal profile of the medium feed rate.

5. CONCLUSION - PART I

The purpose of constructing the mathematical model of a bioreactor behaviour or of a larger section of a bioprocess from theoretical and empirical knowledge is to predict the behaviour or performance of the system. This can be done in a computer simulation and it can be used for different purposes. In general, it is usually the optimization of the process by a selected process aspect (productivity and the best economy are usually the most often used ultimate criteria). Apart from process simulation, an appropriately modified mathematical model of the process can be used in the process control scheme.

The mathematical model of a bio-system should respect the relationships and interactions between the living organism and its environment.

The model should, however, also be as simple as possible and yet adequately accurate in describing the key factors characteristic for the behaviour of the real process.

In the construction of biokinetic models, a knowledge of the relevant metabolic pathway and physiology of the culture is required. Most often a simplified scheme of reactions involved in the process is characterized by:

- reaction stoichiometry
- reaction rates.

The reaction rates are usually approximated by using the relationships summarized in Table 2.1 derived from the theory of enzyme and chemical reactions. The relevant knowledge of (microbial) physiology usually involved quantitative assessment of interactions between the process variables, e.g. the inhibitory effect of products on the cellular growth, and of understanding the response of the microbial culture to its environment, e.g. pH, shear stress, and/or conditions changing the behaviour of the cell, etc.

Typically, the construction of mathematical models for bio-processing, also as discussed in this text, should follow the algorithm suggested in Figure 5.1.

Mass balances and well established methodologies of expressing them in the mathematical form may not necessarily be a familiar concept for non-engineers. The principles of mass balancing have been included in a somewhat abbreviated form as relevant for the purpose of this text. It is easily noted that there are many different variations of the basic mass balance equation

$$Input = Output \pm Accumulation$$

It takes a certain amount of practice in order to be able to handle this basic mass balance scheme with comfortable confidence. Since the mass balances in the context of this book are most often developed for the more or less well defined bioreactor system, the basic reactor engineering concepts of the perfectly mixed and the plug flow reactors have been briefly introduced.

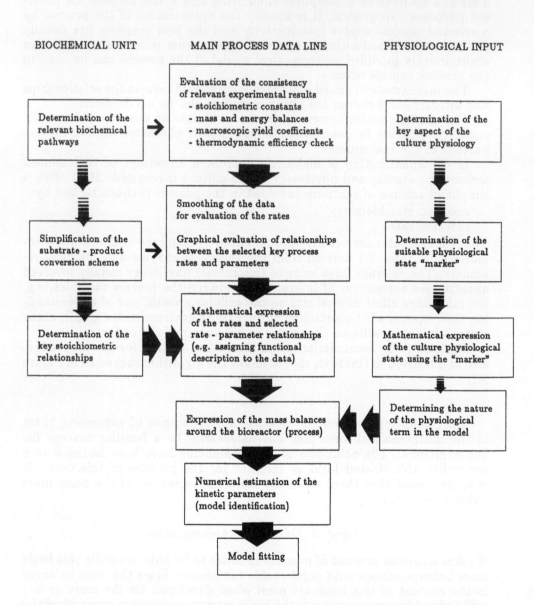

BIOCHEMICAL UNIT   MAIN PROCESS DATA LINE   PHYSIOLOGICAL INPUT

Determination of the relevant biochemical pathways

Evaluation of the consistency of relevant experimental results
- stoichiometric constants
- mass and energy balances
- macroscopic yield coefficients
- thermodynamic efficiency check

Determination of the key aspect of the culture physiology

Simplification of the substrate - product conversion scheme

Smoothing of the data for evaluation of the rates

Graphical evaluation of relationships between the selected key process rates and parameters

Determination of the suitable physiological state "marker"

Determination of the key stoichiometric relationships

Mathematical expression of the rates and selected rate - parameter relationships (e.g. assigning functional description to the data)

Mathematical expression of the culture physiological state using the "marker"

Expression of the mass balances around the bioreactor (process)

Determining the nature of the physiological term in the model

Numerical estimation of the kinetic parameters (model identification)

Model fitting

Figure 5.1: The scheme of constructing the mathematical model of a fermentation process.

Apart from the process material balancing aspects, mass and energy balances for fermentation processes are used also as a preparative step for the mathematical model construction in order to:
- check the consistency of the experimental results;
- estimate the yield coefficients for the production of biomass, utilization of substrate and formation of products;
- estimate the rates of oxygen and substrate(s) uptake in the process;
- assess the efficiency of substrate energy recovery in the products;
- estimate the rate and amount of heat evolution in the process;
- develop relationships between the utilization of substrate and the production of biomass and other organic materials;
- provide a tool for indirect estimate of the product quantities when direct measurements are not possible;
- provide the backbone for the construction of mathematical models representing the process.

Minkevich and Eroshin[48] and Erickson et al.[14] introduced in their work a useful concept of deriving mass and energy balances for the microbial process which are based on the number of available electrons assuming that:
- the number of available electrons per carbon atom in biomass is constant,
- the weight fraction of carbon in biomass is relatively constant,
- the heat of reaction per electron transferred to oxygen for a wide variety of organic molecules is constant.

A stoichiometric equation with respect to growth, carbon source utilization, non-cellular product formation, oxygen uptake, carbon dioxide evolution, etc. can be regarded as the basis of the law of conservation of substrate metabolized by the microorganism. For the expression of useful macroscopic yield coefficients the weight fraction of carbon in the organic matter has to be defined. Biomass energetic yield coefficient (η) can be estimated from several different relationships involving the substrate, different biomass yield expressions, nitrogen consumption, carbon dioxide production, and oxygen consumption, respectively. Agreement of these estimates from two or more of these relationships reflects the accuracy of the material and energy balances drawn. If the estimate differences are too large this may be an indication of inconsistences perhaps because of:
- formation of products other than those considered,
- an inaccurate assumption such as the nitrogen content of the biomass,
- measurement errors.

The balances thus serve as a valuable tool for the assessment of consistency of the experimental data.

Smoothing of experimental data by approximating the results by a continuous and differentiable function can conveniently be done, usually by using the "spline" or "piecewise polynomial" functions. This approach is greatly facilitated by the use of readily available computer programs. A simple block diagram in Figure 7.2 outlines the procedure for smoothing of the experimental data leading to the estimation of relevant fermentation rates.

The development of relationships for expressing the rates of metabolic reactions in the fermentation process usually relies heavily on the theory of enzymatic or chemical reactions. There is a limited selection of mathematical functions used to express the rates of fermentation process variables. The expression of these rates is usually the most crucial aspect in constructing the mathematical model which may be relying heavily on the system mass balances and their terms with rate expressions in them. On the other hand, this modeling approach tends to include and reflect the actual process variables and thus represents the real process in very practical terms. The model of this sort constitutes potentially a very powerful process analysis tool. When constructed, the model is only a "general model" consisting of a number of equations including a number of model parameters, coefficients, which need to be numerically "identified". At this point it is crucial for the model to be capable of reflecting the actual process behaviour.

Identification of the model parameters is done by fitting the model to the actual process data. Methods of non-linear and linear regression are usually used for this purpose. These methods, based on mathematical statistics, have become routine and a selection of computerized methods is available and could be applied, depending on the type of the model constructed. A block diagram of a parameter estimation algorithm is presented in Figure 3.2.

When the model parameters are numerically established, the model is "identified" and could be applied in a number of different process simulation assignments:

- to predict the process behaviour,
- to study the process behaviour in different process operating alternatives such as the change from batch to fed-batch, continuous or immobilized-cell modes,
- to assess the bioreactor stability and steady state operation,
- to test the process performance under new or unconventional conditions,
- to test the process for scale-up procedures,
- to optimize the process performance,
- to establish the process control algorithms,
- to use the model in the process control,
- to use the model as part of the process economics estimation where it may represent one of the "blocks" in the overall process "block diagram".

Most of the process performance simulation studies are done following procedures established in studies of chemical reactors.

These is an impressive accumulation of know-how in this area which still waits to be applied in bio-processes.

If the model is more structured, including the biochemical and physiological culture aspects reflected by properly selected "marker terms" it may reflect the real bioprocess very well indeed. Of course, including a relatively not so well defined concept of the "culture physiological state" in a mathematical model of its performance could also have potential "pitfalls".

On the other hand, it has been demonstrated recently[90-93] that this approach may lead to new insights into the culture behaviour and its interpretation. In other words, structured models, including interpretable physiological terms, may serve as a powerful tool for further elucidation and "quantification" of microbial physiology. It can be expected that this approach to bioprocess modeling may even serve a useful purpose as a "diagnostic tool"[92] to assist in explanation of certain aspects of microbial culture behaviour[93].

NOMENCLATURE - PART I

Lower case symbols:

$a; b; c; ... z$	coefficients
a	specific interface surface area
d	fungal pellet diameter
$f( )$	a function
$f(\xi)$	a weight function representing the age structure of the culture population[56]
k	constant, rate constant
k_L	mass transfer coefficient (on the liquid side)
ℓ	distance
p	weighing coefficient[5]
q	specific rate of accumulation
$q(S)$	metabolic activity of an individual cell as function of substrate concentration[56]
r	reaction time
s	variation of the residue error in Eq. (3.5)[4]
t	time
y	measured quantity
w_{ij}	weight coefficient in Eq. (3.3)

Upper case symbols:

C	concentration
ΔC	concentration difference
F	feed flow rate; distribution
F_1	mass transfer rate between no-flow and flow-through regions[73]
K	constant, coefficient
O	oxygen
P	product (concentration)
Q	fermentation heat[14,48]
RNA	ribonucleic acid
S	substrate (concentration)
$Q(t)$	active metabolic functional[56]
V	volume
X	biomass concentration
Y	yield coefficient

Greek Symbols:

α	flow-through volume of a batch; (probability)
β	degree of reducibility[14,48]
Δ	a finite interval or a difference; (mean) deviation in Eq.(3.4)[4]
ε	experimental error
η	energetic yield coefficient for biomass[14,48]
λ	statistics in Eq.(3.8)
μ	specific growth rate
ν	degrees of freedom
ξ	chemical energy fraction[14,48]
σ	weight fraction of carbon[14,48]
τ	time
δ	partial derivative

Subscripts (lower case):

c	refering to or based on carbon (yield; 1 gramatom)[14]
f	final (Volume)
i	number of elements; initial (Volume)
m	medium
o	initial (or feed)
l; m; n; p; q; r; s; t	indices for a number of atoms in a general chemical molecular formula[14]

Subscripts (capital)

A	acetone
B	biomass, butanol
BA	butyric acid
L	liquid
O	oxygen
P	product
R	aerated liquid
S	substrate
E/S	ethanol on substrate (yield)
P/S	product on substrate (yield)
V	volume
X	biomass
X/O	biomass on oxygen (yield)
X/S	biomass on substrate (yield)

Superscripts

m	number of measured experimental points, Eq. (3.3)
n	(general) number of steps or intervals; number of unknowns in Eq. (3.3)
N	(general) number of steps or intervals (also subscripts)[5]
★	equilibrium at the interface (concentration)
bar	mean value

REFERENCES FOR PART I

1. Aris, R. 1965. *Introduction to the Analysis of Chemical Reactors*. Prentice Hall, Englewood Cliffs, N.J.
2. Atkinson, B. 1974. *Biochemical Reactors*. Pion Press, London, U.K.
3. Bailley, J.E., Ollis, D.F. 1977. *Biochemical Engineering Fundamentals*. McGraw Hill, N.Y.
4. Bard, Y. 1974. *Nonlinear Parameter Estimation*. Academic Press, N.Y.
5. de Boor, C. 1978. *A Practical Guide to Splines*. Springer Verlag, N.Y.
6. Bradley, L.P., Hax, A.C., Magnashi, T.L. 1977. *Applied Mathematical Programming*. Addison-Wesley, Reading, Massachusetts.
7. Burkowski, F.J. 1973. Comm. ACM **16**: 635.
8. Burlirsch, R. 1966. Num. Math. **8**: 1.
9. Christiansen, J. 1970. Num. Math. **14**: 399.
10. Conley, W. 1981. *Optimization: A Simplified Approach*. Petrocelli Book, N.Y.
11. Dew, P.M., Walsh, J.E. 1981. ACM Trans. Math. Soft. **7**: 295.
12. Dixon, L.C.W. 1973. *Nonlinear Optimization*. English Univ. Press, London, U.K.
13. Duris, C. 1980. ACM Trans. Math. Soft. **6**: 92.
14. Erickson, L.E., Minkevich, I.A., Eroshin, V.K. 1978. Biotechnol. Bioeng. **20**: 1595.
15. Fan, L. 1966. *Continuous Maximum Principle*. J. Wiley, N.Y.
16. Frederickson, A.G., Ramkrishna, D., Tsuchiya, H.M. 1967. Math. Biosci. **1**: 327.
17. Frederickson, A.G., Tsuchia, H.M. 1977. Microbial Kinetics and Dynamics, in *Chemical Reactor Theory*, Lapidus, L., Amundson, N.R. Eds. Prentice Hall Inc. Englewood Cliffs, N.J., p. 405.
18. Frome, E.L. 1974. Appl. Stat. **23**: 238.
19. Gear, C.W. 1971. Comm. ACM **14**: 185.
20. George, R. 1963. Comm. ACM **6**: 663
21. Gill, P.E., Murray, W., Wright, M.K. 1981. *Practical Optimization*. Academic Press, N.Y.
22. Gottfried, B.S., Weismann, J. 1973. *Introduction to Optimization Theory*. Prentice Hall, Englewood Cliffs, N.J.
23. Grimes, *et al.* 1978. ITPACK 2A: *Report 164*, Center for Num. Anal., Univ. Texas, Austin, TX.
24. Haken, H. 1977. *Synergetics*. Springer Verlag, Berlin, Germany.
25. Harder, A., Roels, J.A. 1982. Application of Simple Structured Models in Bioengineering, *Adv. Biochem. Eng. 21, Microbes and Engineering Aspects*. Springer Verlag, Berlin, Germany, p. 56.

26. Hartmann, K., Slinko, M.G. 1972. *Methoden und Programme zur Berechnung Chemischer Reaktoren.* Akademie Verlag, Berlin, Germany.

27. Havlik, I., Janik, P., Votruba, J. 1984. Collect. Czech. Chem. Commun., **49(2)**:386.

28. Hiebert, K.L. 1981. ACM Trans. Math. Soft. **7**: 1.

29. Hiebert, K.L. 1982. ACM Trans. Math. Soft. **8**: 5.

30. Himmelblau, D. 1970. *Process Analysis by Statistical Methods.* J. Wiley, N.Y.

31. Himmelblau, D. 1972. *Applied Nonlinear Programming.* McGraw Hill, N.Y.

32. Iserman, R. 1981. *Digital Control Systems.* Springer Verlag, N.Y.

33. Jeffreson, T.H. 1973. TJMARI: Sandia Tech. Rep. No. SLL73-0303.

34. Kafarov, V.V., Dorochov, J.N. 1976. *System Analysis of Processes in Chemical Technology: Principles of Strategy.* Nauka, Moscow, USSR.

35. Kafarov, V.V., Dorochov, J.N., Lipatov, L.N. 1982. *Systems Analysis of Processes in Chemical Technology: Statistical Methods of Process Identification.* Nauka, Moscow, USSR.

36. Kamien. M.I., Schward, P. 1981. Dynamic Optimization, *The Calculus of Variation and Optimal Control in Economics and Management*, Vol. 4. North Holland, N.Y.

37. Kaplan, E.L. 1982. *Mathematical Programming and Games.* J. Wiley, N.Y.

38. Kaupe, A.F. 1963. Comm. ACM **6**: 313.

39. Kuester, J.L., Mize, J.H. 1973. *Optimization Techniques in FORTRAN.* McGraw Hill, N.Y.

40. Leaf, G. 1977. DISPL: A software package for one and two spatially dimensional kinetics diffusion problems. *Report ANL-77-12*, Nat. Energy Software Center, Argonne National Labs, Argonne, IL.

41. Levenspiel, O. 1962. *Chemical Reaction Engineering*, J. Wiley, N.Y.

42. Machura, M. 1973. Comm. ACM **16**: 482.

43. Machura M., Sweet, R. 1980. ACM Trans. Math. Soft. **6**: 461.

44. Madsen, N.K., Sincovec, R.F. 1979. ACM Trans. Math. Soft. **5**: 326.

45. Malek, I. 1976. Physiological State of Continuously Grown Microbial Cultures, in *Continuous Culture 6*, Ch. 2, Dean, A.C.R., Ellwood, D.C., Evans, C.G. Eds. Ellis Horwood Ltd., Sussex, England. p. 31.

46. Melgaard, D.K., Sincovec, R.F. 1981. ACM Trans. Math. Soft. **7**: 106.

47. Melgaard, D.K., Sincovec, R.F. 1981. ACM Trans. Math. Soft. **7**: 216.

48. Minkevich, I.A., Eroshin, V.K. 1973. Folia Microbiologica **18**: 376.

49. Naur, P. 1960. Comm. ACM **3**: 312.

50. Nicolis, G., Prigogin, I. 1977. *Self-Organization in Non-Equilibrium Systems*, J. Wiley, N.Y.

51. O'Connell, M.J. 1974. Com. Phys. Comm. **8**: 56.

52. Ollis, D.F. 1947. Ch. 8. Biological Reactor Systems, in *Chemical Theory*, Lapidus, L., Amundson, N.R. Eds. Prentice Hall, Englewood Cliffs, N.J. p. 484.

53. O'Neill, R. 1971. Appl. Stat. **20**: 338.

54. Pazoutova, S., Votruba, J., Rehacek, Z. 1981. Biotechnol. Bioeng. **23**: 2831.

55. Peck, J.E.L. 1962. Comm. ACM **5**: 44.

56. Powell, E.O. 1968. Transient Changes in the Growth Rate of Microorganisms, in *Proc. 4th Symposium Continuous Cultivation of Microorganisms*, Malek, I. Ed., Prague, Academia. p. 275.

57. Ramkrishna, D. 1979. Adv. Biochem. Eng. **11**: 1.

58. Rastrigin, L.A., Manzharov, R. 1977. *Introduction to the Identification of Control Objectives*, Energija, Moscow.

59. Reinsch, C. 1967. Num. Math **10**: 182.

60. Ricica, J., Votruba, J. 1982. Physiological Aspects In Development of Mathematical Models, in *Proc. FEMS Symp. Overproduction of Microbial Products*, Krumphanzl, V., Sikyta, B., Vanek, Z. Eds. Academic Press, London. p. 675.

61. Richardson, J.A. 1973. Comm. ACM **16**: 487.

62. Roberts, S.D. 1982. Teaching Simulation to Undergraduates, in *Proc. 1982 Winter Simulation Conference*, Highland, H.G., Chao, Y.W., Madrigal, O. Eds. IEEE, N.Y.

63. Rodriguez-Gil, F. 1963. Comm. ACM **6**: 387.

64. Roels, J.A. 1980. Biotechnol. Bioeng. **22**: 2457.

65. Sammet, J. 1976. Comm ACM **19**: 656.

66. Sargent, R.G. 1978. Introduction to Simulation Languages, in *Proc. 1978 Winter Simulation Conference*, Highland,H.J., Nielsen, N.R., Hull, L.G. Eds. IEEE, N.Y.

67. Schiesser, W.E. 19812. DSS/2. Lehigh Univ., Bethlehem, PA.

68. Schwefel, H.P. 1981. *Numerical Optimization of Computer Models*. J. Wiley, N.Y.

69. Scientific Computing Consulting Services. 1979. PDEPACK. Colorado Springs, CO.

70. Shanno, D.F. 1976. Trans. ACM Math. Soft. **2**:87.

71. Shannon, R.E. 1976. *System Simulation: The Art and Science*. Prentice Hall, Englewood Cliffs, N.J.

72. Sill, A.L. 1970. Comp. J. **13**: 111.

73. Sinclair, C.G., Brown, D.E. 1981. Biotechnol. Bioeng. **23**: 2831.

74. Snoll, L.E. 1979. *Physico-Chemical Factors of Biological Evolution*. Nauka, Moscow, USSR.

75. Sobotka, M., Votruba, J., Prokop, A. 1981. Acta Biotechnologica 1: 3.

76. Sobotka, M., Votruba, J., Prokop, A. 1981. Biotechnol. Bioeng. 23: 1193.

77. Sobotka, M., Prokop, A., Dunn, I.J., Einsele, A. 1982. Ch. 5. Review of Methods for the Measurement of Oxygen Transfer in Microbial Systems, in *Annual reports on Fermentation Processes, 5.* Academic Press, N.Y.

78. Sobotka, M., Votruba, J., Krumphanzl, V., Hradec, Y. 1982. Kvasny Prumysl (Czech.) 28(4):87.

79. Spath, H. 1968. *Algorithmen zur Knostruktion Glatten Kurven und Flachen,* Odenburg Verlag, Munchen, W. Germany.

80. Synge, M.J. 1963. Comm. ACM 6: 313.

81. Tendler, J.M., Bickart, T.A., Picel, Z. 1978. ACM Trans. Math. Soft 4: 399.

82. Thom, R. 1972. *Stabilité Structurelle et Morphogénèse.* Benjamin, N.Y.

83. Thomas, D.G. 1977. Appl. Stat. 21: 103.

84. Thompson, J.M. 1982. *Instabilities and Catastrophes in Science and Engineering,* J. Wiley, N.Y. p. 111.

85. Votruba, J. 1982. Acta Biotechnologica 2: 119.

86. Votruba, J. 1983. *STIFFSOLVER 81 Manual for the Program,* Archive of Programs CSAV; Pod Vodarenskou vezi 5, Praha 8; Czechoslovakia.

87. Wasscher, E.J. 1963. Comm. ACM 6: 517.

88. Wen, C.Y., Fan, L.T. 1975. *Models for Flow Systems and Chemical Reactors.* Marcel Dekker, N.Y.

89. Whitley, V.W. 1962. Comm. ACM 5: 550.

90. Yerushalmi, L., Volesky, B. 1986. Can. J. Chem. Eng. 64: 607.

91. Yerushalmi, L., Volesky, B., Votruba, J. 1986. Biotechnol. Bioeng. 28: 1334.

92. Yerushalmi, L., Volesky, B., Votruba, J. 1988. Appl. Microbiol. Biotechnol. 29: 186.

93. Yerushalmi, L., Volesky, B., Votruba, J., Molnar, L. 1989. Appl. Microbiol. Biotechnol. 30: 460.

NOTE:

Communication, collaboration .... is always necessary, particularly for getting ahead and solving tasks in interdisciplinary fields - such as biotechnology ....

PART II

FUNDAMENTALS OF MASS BALANCING

6.   MASS BALANCES

In 1789 Antoine Lavoisier concluded from his experiments on combustion
that

> *"....in all operations of art and nature, nothing is created;
> an equal quantity of matter exists both before and after the
> experiment".*

His experiments laid to rest the phlogiston theory and provided the
basis for a quantitative treatment of chemical equations. Without this
breakthrough the development of modern chemical technology would have
been impossible. The simple statement that matter is neither created nor
destroyed was subsequently shown by Albert Einstein to be only approximately true. His famous relation

$$\Delta E \qquad = \qquad -(\Delta m) c^2 \tag{6.1}$$

gives the amount of energy released upon annihilation of an amount of mass
Δm. The minus sign is required because of the definition

$$\Delta m = (final \quad mass) - (initial \quad mass) \tag{6.2}$$

The quantity c is the speed of light. If Δm is measured in *kg* and c in *m/sec*,
ΔE is in *joules*.

For engineering and biotechnology process calculations, it is appropriate
to use the simple statement of Lavoisier rather than the more complicated,
but more correct, relativity theory which is appropriate for nuclear rearrangements where the transformation of matter and energy must be taken
into account. For all physical, chemical and biochemical processes, the
Lavoisier's simple statement is correct.

System

The simple statement that matter is neither created nor destroyed is not in a form suitable for application. To make use of it, a different concept of a system has to be introduced first.

The system used in the considerations herein is the same as that employed in thermodynamics, i.e., it is the portion of the universe on which we concentrate our attention. The term *system* will be used for a designated region which the mass enters or leaves. We will not distinguish between open systems, closed systems, isolated systems or control volumes. Once we have singled out a region for study, we can apply Lavoisier's statement that matter is neither created nor destroyed inside the system after we have chosen a basis.

Basis

We must consider the time period over which we will observe the system. For example, we may decide that we will observe the system for a period of one hour or one week, one second or 17.6 minutes. This is the *basis*. Once this has been decided, we can use Lavoisier's statement in a quantitative way.

In most problems a time basis such as one hour, minute or second is not convenient because input and output will then be inconvenient numbers. We will frequently choose a convenient mass of input or output. This may correspond to a fractional time basis.

Conservation of Mass

After choosing a system and basis we can sketch the general situation as follows:

$$INPUT \longrightarrow \boxed{SYSTEM} \longrightarrow OUTPUT$$

where:

INPUT denotes the total input of mass to the system from all sources during the time period chosen as the basis.

OUTPUT denotes the total output from the system in all streams during the time period chosen as the basis.

The statement of the Law of Conservation of mass is

$$INPUT \quad - \quad OUTPUT \quad = \quad ACCUMULATION \qquad \text{(6.3)}$$

where
> *ACCUMULATION* denotes the mass within the system finally less the mass within the system initially. The final time is equal to the initial time plus the basis time.

Note that the *ACCUMULATION* is an algebraic quantity (it has a sign) while *INPUT* and *OUTPUT* are taken as positive, i.e., the minus sign in Equation (6.3) automatically accounts for the direction of flow once we have identified a stream as input or output. Since we will not consider nuclear rearrangements, the conservation of mass applies to each element involved in the process. Equation (6.3) therefore states the conservation law for any element in the process. If no chemical reactions occur within the system, then each chemical species is also conserved and Equation (6.3) is the conservation statement for species. For mass balances with no chemical reactions, the species form will be used. The procedure will be more complex when chemical reactions occur. However, in both cases, the overall mass balance applies.

Problem Solving Technique

The following steps should become part of the approach taken in writing mass balances and solving them for the desired unknowns.
1. *Draw a process flowsheet* and label with the information available. Be sure the operations in the process and the relationship between the operations are clearly understood. Identify the species which appear in each stream. Restate the unknowns in terms of flows or compositions on the flowsheet. These steps will help make clear exactly what the problem is.
2. *Choose a system.* Draw an envelope around those parts of the process which are to be examined. The process streams *cut* by this envelope are the inputs and outputs for the system considered.
3. *Pick a basis.* This sets the scale of the calculations. It is most important because numbers which are not on the same basis cannot be used in the same equation except as ratios.
4. *Make the calculations.*
5. *Check the answers.* The units have to be consistent and the numerical values *reasonable* (flow rates > 0, mole fractions < 1, etc.)

In most cases judging *rightness* of numerical mass balance calculations comes with experience which can hardly be taught theoretically. The most obvious check of the numerical calculations is to see whether the mass balances *close*. Otherwise, the assumptions and the structure of the balance equation(s) needs to be checked.

Material Balances as Data Checks

When analyzing an existing process, more flow rate and composition data are often available than needed for a complete mathematical specification of all the material balances. In this situation material balances may be used to check the reliability of the given information. If the input does not equal the output it is said that the material balance does not *close*.

If the material balance does not close to sufficient accuracy, corrective action must be taken. The action depends on the process and the purpose for which the calculations are being made. Often some of the data are suspect, e.g. certain analyses are more difficult than others, etc. In these cases the suspect data are not used and the superfluous material balances are used to supply the (now) missing information.

6.1   SYSTEMS WITHOUT CHEMICAL REACTIONS

Classification of Problems

This Section will address problems without chemical reactions The problems analyzed with the Law of Conservation of Mass will fall into one of three categories:

A. *Steady State Operation* - for processes which operate at steady state the accumulation is zero. Equation (6.3) simplifies to

$$INPUT \quad = \quad OUTPUT \qquad (6.4)$$

Since many continuous processes operate at nearly steady state, the simplified form is often used. In process design work steady state operation is always assumed in developing the process flowsheet.

The other two categories are often used for batch processes and for continuous processes where flows vary with time.

B. *Intermittent Operation* - for processes which operate with intermittent flows there is an accumulation. Equation (6.3) must be used. For processes classified as intermittent we often know total input, or output or accumulation during the time interval under study and Equation (6.3) yields an algebraic relationship.

C. *Transient Operation* - for processes which operate with time varying flows Equation (6.3) must be used in the form of a differential Equation. In transient processes the flows must be known as functions of time.

Further, the processes in any one of these categories can be classed as:

1. physical operations - processes in which no chemical reactions occur
2. chemical operations - processes in which some species undergo chemical reaction

The strategy used for solving problems in each of these classes is slightly different, therefore, one must know whether the process under study contains only physical operations or a combination of physical and chemical operations. We will first look at physical operations.

6.1A   STEADY STATE PROCESSES WITHOUT CHEMICAL REACTIONS

In all processes the total mass is conserved, but in purely physical processes the mass of each species is also conserved. Therefore, the Conservation of Mass Law for each species can be written as well as for the total mass. However, one of these equations will not be independent because it is the one which concerns the species which can be determined by the difference from the total sum.

Not every equation which can be written expressing conservation of mass will be independent. In a system where no chemical reactions occur, the number of *independent* mass balances is equal to the number of species present, N_{sp}.

The ease of numerical solution is affected by the set of independent equations we use. In general, simpler, faster manual solutions are possible if the equations used are the overall mass balance and the $(N_{sp} - 1)$ species balances which contain the fewest unknowns. In more complicated problems, it will pay to use equations with fewer unknowns. The basis is usually chosen so that the percentage is numerically equal to the mass of each entering species. This saves one multiplication for each species in return for a single multiplication (proration) at the end of the problem. This may be a substantial saving of effort if a large number of species is present.

In a problem involving gas mixtures a basis of 100 moles could be chosen to take advantage of the fact that gas compositions are in mole percent.

Species which appear in only two streams of a process considered are called *tie substances*. A little time spent looking for tie substances will often be repaid by fast, simple problem solutions. In a steady state process without chemical reaction each species must appear in at least one inlet and one outlet stream, therefore, a tie substance is one that appears in the smallest possible number of streams. The search for tie substances is merely the first step in the search for the equations containing the fewest unknowns.

A word of caution is in order with regard to the use of tie substances. One should not use as tie substances species which are present in trace quantities given with questionable accuracy. A compound whose analysis is given to only one significant figure would be a poor choice for a tie substance. If there are several tie substances, those more accurately analyzed should be used.

For complex processes there will be a large number of independent material balances. The number of independent systems in a process is equal to the number of operations in the process. The stream splitters $(\longrightarrow^{\uparrow}\longrightarrow)$ and mixers $(\longrightarrow_{\downarrow}\longrightarrow)$ are counted as operations. In complex processes the choice of system can influence the ease of solution of a problem in a fashion analogous to that for the choice of basis.

Depending upon the data available, the solution may be sequential, i.e., one solves first one system then another. Therefore, it is advantageous to choose as the first system the one for which there is the most data. The complexity of the problem can vary depending upon which pieces of information are given. It will pay dividends to analyze the problem to set an overall strategy before an attempt is made to solve it in detail.

In some systems a *bypass process stream* is encountered. This is often the case when a sub-process will perform better than required. In order to avoid "product quality give away" and meet product specifications, some of the feed may be bypassed around the unit and mixed with the product. In general, it can be postulated that balances around the entire process provide no information on bypass (and/or recycle streams). No information can be obtained about streams not cut by the system boundary. Balances around mixing points are particularly useful for processes involving bypass (and/or recycle).

A careful choice of the more appropriate mass balance equations from all those which can be written can greatly facilitate the solution. A certain amount of experience is helpful for making the choice which can only come with more practice.

6.1B   INTERMITTENT OPERATION WITHOUT CHEMICAL REACTIONS

The system analyst is frequently concerned with processes in which Accumulation is important. The importance of the accumulation within the system may be judged in relation to the input and output by the material balance

$$INPUT \quad - \quad OUTPUT \quad = \quad ACCUMULATION$$

Even though no process operates at an exactly steady state, it is always possible to pick as a basis a time sufficiently long so that the accumulation is small relative to the total input and output. This is possible because the maximum accumulation is equal to the maximum amount of mass the process equipment could hold. If the time period is sufficiently long, the total throughput can be made much larger than this maximum accumulation. This assumption is in effect made when analyzing processes by the steady state balance as outlined in the previous section. The compositions and flow rates for steady state systems are really average values over long times.

In some cases, however, one may be interested in changing the conditions in a process and observing the change in product quality. In an experimental situation like this, since the plant cannot take the risk of making an unsaleable product (if the changes reduce product quality), the tests may be run over a time period so short that accumulation is important. In using the complete mass balance equation the analyst must have information on the average compositions and flows over the time period of the test. Using these values the resulting mass balance equations are algebraic.

With regard to the systems operating in an intermittent manner, it is helpful to realize that if the composition of the streams entering and leaving the system are equal only one independent material balance can be written.

Great care must be exercised when determining the sign in calculating the accumulation. A negative accumulation indicates that there is less mass inside the system at the end than there was at the beginning.

EXAMPLE 6.1 (Section 6.1B)

The dynamics of the nonideal flow pattern of fluid in large scale fermentors can be described by using combined models of flow. For description of the fluid flow in an anaerobic wastewater treatment plant associated with beet sugar production a mathematical model of mixing was developed by Heertjes[6]. The liquid entering the anaerobic bioreactor is divided into the blanket bed section stream and the stream feeding the sludge bed section. The sludge bed section is partially mixed, partially coupled with a dead space. Both sections (blanket and sludge bed) are interconnected with fluid flows. The liquid leaves from the blanket section. For identification of the liquid flow pattern the residence time technique can be applied. The time course of a tracer is used to estimate the flow structure and ratio. Write the mass balance for the tracer in the sludge bed and in the blanket sections.

Solution:

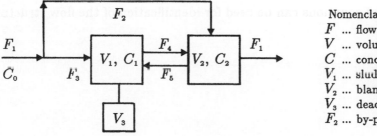

Nomenclature used:
F ... flow
V ... volumes
C ... concentrations
V_1 ... sludge bed section
V_2 ... blanket section
V_3 ... dead space
F_2 ... by-pass flow

System: Volume of apparatus.

Mass balances for the volumetric flow and concentration have to be written.

$$INPUT - OUTPUT = ACCUMULATION$$

The flows are in steady states. For the bypass splitter the flow balance is:

$$F_1 = F_2 + F_3$$

and

$$V = V_1 + V_2 + V_3$$

$$F_3 \cdot C_0 - F_4 \cdot C_1 + F_5 \cdot C_2 = \frac{d(V_1 C_1)}{dt}$$

$$F_2 \cdot C_0 + F_4 \cdot C_1 - F_1 \cdot C_2 - F_5 \cdot C_2 = \frac{d(V_2 C_2)}{dt}$$

After rearrangement we receive the differential equations for the time course of the tracer in the blanket and the sludge bed sections:

$$\frac{dC_1}{dt} = (F_3 C_0 - F_4 C_1 + F_5 C_2) / V_1$$

$$\frac{dC_2}{dt} = (F_2 C_0 + F_4 C_1 - F_1 C_2 - F_5 C_2) / V_2$$

This set of equations can be used for identification of the flow structure.

6.2   SYSTEMS WITH CHEMICAL REACTIONS

Consider a system in which a chemical reaction occurs

$$Input \longrightarrow \boxed{SYSTEM} \longrightarrow Output$$

The statement of Lavoisier allows writing the following equation for the total mass

$$(Input\ of\ mass)\ -\ (Output\ of\ mass)\ =\ (Accumulation\ of\ mass)$$

If reactions occur the above equation cannot be written for each species. A balance, however, can be written for each entity which is not destroyed in the reactions. Since elements are not destroyed in chemical reactions, we can write the basic above equation as an element balance

$$\left|\begin{matrix} Input\ of \\ element\ j \end{matrix}\right| - \left|\begin{matrix} Output\ of \\ element\ j \end{matrix}\right| = \left|\begin{matrix} Accumulation\ of \\ element\ j \end{matrix}\right| \qquad (6.5)$$

We can also write a balance similar to Equation (6.5) for any group of elements which is not split apart or formed in the particular system we are studying. In addition, we can also write a balance for each species which does not take part in the reaction. Species which do not enter into a reaction are called *inerts*. Therefore, we can make a species balance for each inert.

$$\left|\begin{matrix} Input\ of \\ inert\ species\ i \end{matrix}\right| - \left|\begin{matrix} Output\ of \\ inert\ species\ i \end{matrix}\right| = \left|\begin{matrix} Accumulation\ of \\ inert\ species\ i \end{matrix}\right| \qquad (6.6)$$

Chemical Reaction Process Definitions

Once the number of independent chemical reactions is determined, it is usually a simple matter to write the required number of equations involving all of the species. Three terms frequently used to specify the performance of a chemical or biochemical reactor should be defined here:

Yield – the ratio of the moles of limiting reactant consumed to make the desired product to the moles of limiting reactant fed. Sometimes the words *Conversion to* are used instead of yield.

Conversion – the ratio of the moles of limiting reactant consumed (to make any product) to the moles of limiting reactant fed.

Selectivity –   the ratio of the moles of limiting reactant consumed to make the desired product to the total moles of limiting reactant consumed. It is equal to the ratio of Yield to Conversion as defined above.

From these definitions it is easy to see that $Yield \leq Conversion$ with the equality holding when there is only one reaction.

Also, $0 \leq Selectivity \leq 1.0$

Note that the limiting reactant is that reactant which would disappear first if we could make the reaction go as far toward completion as possible, i.e., it is the reactant present in the smallest stoichiometric amount.

The Number of Independent Material Balances

A general rule for determining the number of independent material balances can be formulated which may be written for any system:

The number of independent material balances equals the number of components present in the system.

For a system in which only physical processes occur the number of components, N_C , equals the number of species, N_{sp}. For a system involving chemical reactions the number of components is equal to the difference between the number of species present and the number of independent chemical reactions. The number of independent chemical reactions is frequently established and known. If it is not, it may be obtained from an *atomic matrix* as follows. The atomic matrix is made up from the molecular formulas of the species entering into the chemical reactions. The entries in each column are the subscripts of each element appearing in the molecule heading the column. For example, for a complete aerobic bioconversion of a hexose sugar $(C_6H_{12}O_6)$ molecule into CO_2 and water according to the stoichiometric formula:

$$C_6H_{12}O_6 \quad + \quad 6\,O_2 \quad \longrightarrow \quad 6\,CO_2 \quad + \quad 6\,H_2O$$

the atomic matrix will be:

	$C_6H_{12}O_6$	CO_2	H_2O	O_2
C	6	1	0	0
H	12	0	2	0
O	6	2	1	2

Inert substances not entering the main reaction are not included in the atomic matrix. The difference between the number of columns in the matrix and the rank of the matrix is equal to the number of independent chemical reactions. For the matrix above we find that the rank is three and therefore, the number of independent reactions if four minus three or one (which we knew *a priori*). For the entire system

$$N_C \quad = \quad 5 - 1 \quad = \quad 4$$

An alternative way of stating this is that the number of components is equal to the number of inerts plus the rank of the atomic matrix:

$$N_C \quad = \quad N_I \quad + \quad r_A \tag{6.7}$$

Here N_I is the number of inert species and r_A is the rank of the atomic matrix made up from the reacting species.

It is instructive to use the atomic matrix method for a problem in which additional information on process chemistry is given. If we have additional information on the process chemistry, it can always be incorporated into the atomic matrix. In any event, we must be sure to use all of the information that is available.

In summary then, we can use the atomic matrix to determine the maximum number of independent reactions which might occur between the species in the system. We then check to see if additional information can be found about the impossibility of some of the reactions. If some occur to a negligible extent, we can use this information to write additional material balances. If we are given the reactions which occur, we use this number in preference to that obtained from the matrix because the former contains the additional chemical information. When analyzing an existing process, the analyst will often have more flow rate and composition data than needed for a complete mathematical specification of all the material balances. In this situation, he may use material balances to check the reliability of his information.

6.2.1   PROCESSES WITH (BIO)CHEMICAL REACTIONS
STEADY STATE SYSTEM WITH CHEMICAL REACTIONS

Since the elements are conserved in a chemical reaction, element balances can be made for the elements contained in all species taking part in the chemical reaction.

The element balances are written on atoms and the inert balance on moles. A mass balance would be obtained by multiplying each equation by the appropriate atomic or molecular weight. Simpler equations result if moles or atoms are balanced.

In general, for steady state systems with chemical reactions, the following rules are useful:

– Species balances can be made for inerts.

– Element balances can be made for the elements contained in the reacting species.

– Element atom balances and species mole balances are to be preferred to mass balances because the latter yield equations with more complicated coefficients.

– When stream quantities are unknown and compositions are known, it is usually best to assign symbols to the unknown quantities and to set up algebraic relations for each element (or species in the case of inerts). The resulting equations will be linear algebraic equations.

– For reacting systems, the overall mass balance is usually not used as one of independent material balances in contrast with the usual strategy for physical systems.

Certain practical advice can be derived from study and solutions of typical reactor-based problems:

– When reactor performance is specified, the best basis is a fixed amount of feed. In this situation, it is also usually best to pick your first system to be the reactor.

– If reactor performance is unknown but the reactor effluent composition is known, it is best to choose a fixed amount of effluent as the basis.

– When reactor performance is unknown, it is usually best to set up algebraic equations.

– When reactor performance is known, it is usually easiest to work with the reactions directly. This, in reality, is doing element balances in our head.

Reaction Processes With Recycle And/Or Purge

For a number of processes, it would be wasteful to operate them in an *open-ended* mode. In a real process, the large quantity of unreacted substance would be separated, mixed with fresh feed and reprocessed in the reactor. This separation and subsequent reuse of raw materials is called *recycle*. A special case of recycle is encountered in fermentation processes whereby microbial cells represent at least one of the process products, functioning as a *self-propagating* catalyst at the same time. It is quite desirable to maintain high concentrations of this *bio-catalyst* in the reactor. One way of accomplishing it is to separate biomass from the fermentation broth and recycle it back into the bio reactor. The products of the fermentation in that case are usually found in the supernatant broth from which they are recovered by different methods.

The product recovery system may include distillation columns, sorption columns, crystallization tanks, evaporators, etc. In most chemical or biochemical processes, the number of pieces of equipment involved in product recovery processes is much greater than the number of chemical reactors. The net effect of employing the recycle in a process is to increase its overall yield. In order to eliminate the accumulation of undesirable substances within the system so that it is possible to maintain product quality, a purge stream is taken from the recycle. This purge is originally meant to remove inert species from the system. Purge streams are frequently subjected to further processing to recover raw materials and products. The system with a purge stream is shown in the schematic diagram below.

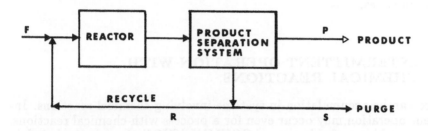

If we take the entire process as the system and write an inert balance for steady state and no inert in the product.

$$F \, x_i \quad = \quad S \, x_R$$

(6.8)

where

$$S \quad = \quad \text{molar flow rate of purge}$$
$$R \quad = \quad \text{mole fraction inert in purge}$$

The purge flow rate is thus set by the rate of input of inert $(F \, x_i)$ and the concentration of inert which can be tolerated in the recycle. The level of an inert which can be tolerated in the system is usually set by the amount of inert which can be tolerated in the reactor. Too much inert material is detrimental as it reduces the concentrations of the other species and thus reduces the rate of reaction.

Balances With Arbitrary Components

Once the number of components has been determined, element balances can be made. Sometimes it is advantageous to use arbitrary components composed of groups of elements which are not split in the overall chemical reaction. Balancing arbitrary components is correct when the arbitrary components are not split in the chemical reactions. The use of arbitrary components is particularly advantageous when there are simple addition reactions in series.

6.2.2   INTERMITTENT OPERATION WITH CHEMICAL REACTIONS

There can be accumulation in systems involving chemical reactions. Intermittent operation may occur even for a process with chemical reactions. In solving problems of this type *ACCUMULATION* must be included in each independent material balance: in this case, in each element balance. The number of independent balances is still equal to the number of components as in steady state operation.

Mass Balances For Chemical Processes With the Equilibrium Restriction

In designing a process, it is useful to have a limiting behaviour for the process. Equilibrium (chemical or physical) provides this limit. For example, in calculating the yield from a chemical reactor the maximum conversion and yield are those corresponding to chemical equilibrium at the temperature and pressure existing at the exit of the reactor. The latter two are particularly limited in their range in the case of biochemical reactors.

Therefore, even if the chemical kinetics were unknown, the limiting performance of the reactor could be calculated by assuming equilibrium in the reactor effluent. The assumption of chemical equilibrium permits process calculations to be made in the absence of detailed kinetic data. How close real reactors perform to this idealization depends upon a number of factors. We only note here that there are processes of industrial importance which use catalysts active enough to yield reaction products nearly in their equilibrium proportions. Each reaction at equilibrium provides a relationship between the compositions of the reactor effluent in addition to the mass balances.

The equilibrium constants for a number of important gaseous reactions are available in different chemical and engineering tables and handbooks. Particularly for biocatalytic reactions important equilibrium limitations exist based for example on established acid-base equilibrium relationships.

EXAMPLE 6.2 (Section 6.2)

The spray drying of yeast *Saccharomyces cerevisiae* was used to produce active dry baker's yeast. After filtration the cake containing 30% solids (10^{10} viable cells per gram of solids) was processed in an Anhydro Laboratory Spray Dryer. In the Table below, data can be found determined for the temperature of 250°C of the incoming hot air stream. (According to Labuza, T.P., LeRoux, Z.J.P., Fan, T.S. 1970. Biotechnol. Bioeng. *12*: 135)[7].

Feed rate [kg/h]	Outlet temp. [°C]	Product moisture [g H_2O/100g solids]	Viable cells [cells/g solid]
2.0	150	1.75	10^3
3.3	125	1.68	4.2×10^4
6.3	90	1.37	6.1×10^5
8.1	70	7.31	6.1×10^7

Assuming that the kinetics of thermal death is of the first order, evaluate the rate constant and the yield of viable cells per kg of product.

Solution:

The denaturation reaction takes place in the dryer which is then functioning like a reactor with a purge.

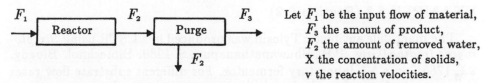

Let F_1 be the input flow of material,
F_3 the amount of product,
F_2 the amount of removed water,
X the concentration of solids,
v the reaction velocities.

The balance of solids and viable cells has to be formulated. Supposing that the system is perfectly mixed and the following reaction takes place:

$$Viable\ Cells \longrightarrow Nonviable\ Cells$$

Then the balances assume the following form:

Mass balance of solids in the purge:

$$INPUT \quad - \quad OUTPUT \quad = \quad 0$$
$$F_1 X_1 \quad - \quad (F_2 X_2 + F_3 X_3) \quad = \quad 0$$

Mass balance of viable cells with a reaction of the 1st order:

$$INPUT \quad - \quad OUTPUT \quad = \quad RATE\ OF\ THERMAL\ DEATH$$
$$F_1 X_1 v_1 \quad - \quad (F_2 X_2 v_2 + F_3 X_3 v_3) \quad = \quad k \cdot v_2$$

Because $X_3 = 0$ the equation for k can be written the following way:

$$k = F_1 X_1 (V_1 - V_2) / v_2.$$

The yield of viable cells is given by:

$$yield = F_2 X_2 v_2 / F_2 = X_2 v_2$$

Where $X_2 = 10^5/(\text{Product moisture g } H_2O + 100)$ [g solid/kg product].

The results are summarized in the table below:

Outlet temp. [°C]	150	125	90	70
k[g solid/h]	6.0×10^{10}	2.36×10^9	3.1×10^8	3.98×10^6
Yield $\left[\frac{Viable\ cells}{kg\ product}\right]$	9.8×10^5	4.1×10^7	6.0×10^8	5.68×10^{10}

EXAMPLE 6.3 (Section 6.2)

The macrolide antibiotic Tylosin was produced in a CSTR using *Streptomyces fradiae* (Gray, P.P., Bhuwapathanapun, S. 1980. Biotechnol. Bioeng. **22**, 1785)[5] in a 5-L laboratory fermentor. For different substrate flow rates the concentrations of product and biomass were measured.

Flow rate [mL/h]	50	100	150	200	350
Biomass conc. [g DW/L]	33	29	28	22	17
Tylosin conc. [g/L]	0.7	0.4	0.2	0.1	0.01

Evaluate the specific growth rate and the specific production rate, using the above experimental data.

Solution:

$F \longrightarrow$ V $F \longrightarrow$

System: content of the tank.

The specific growth and production rates are defined as rates based on unit volume divided by the biomass concentration. Because the system is operating at a steady state condition

$$RATE\ OF\ INPUT \quad = \quad RATE\ OF\ OUTPUT$$

and the rate of input is the rate at which the product species is formed by the reaction.

The rate of output is the rate at which a species leaves the system.

Then both balances for the biomass and the product can be written in the following way:

Mass balance of biomass:

$$V \mu X \quad = \quad F X$$

Mass balance of tylosin:

$$V q X \quad = \quad F P$$

Where μ and q are the specific growth and production rates respectively. After rearrangement of the balances, the specific rates are given by the following equations:

$$\mu = \frac{F}{V} \qquad q = \frac{FP}{VX}$$

The resulting rates are summarized in the Table below:

Flow rate [mL/h]	50	100	150	200	350
μ [1/h]	0.01	0.02	0.03	0.04	0.07
$q \times 10^3$ [1/h]	0.212	0.2758	0.2142	0.1818	0.0411

<u>EXAMPLE 6.4</u> (Section 6.2)

The growth of *Methylomonas* L3 on methanol was observed in a CSTR. Working volume of the tank was $V = 0.355$ (Hirt, W. 1981. Biotechnol. Bioeng. **23**, 235)[7]. The concentration of methanol in fresh medium was 1 g/L. The results of the experiment at a steady state are summarized in the following Table:

F Feed rate [mL/h]	67	70	120	155	165
X Biomass conc. [g DW/L]	0.441	0.47	0.527	0.563	0.563
M Methanol conc. [mg/L] in output stream	30	20	23	16	7

Evaluate the $Y_{s/x}$ and the specific growth rate.

Solution:

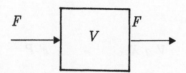

System: content of the tank at steady state.

The specific growth rate is defined as the biomass production rate based on the unit of volume and divided by the biomass concentration.

The macroscopic yield coefficient is the rate of substrate uptake divided by the growth rate. Because the system is operating at a steady state for both substrate and biomass,

$$RATE\ OF\ INPUT = RATE\ OF\ OUTPUT$$

then the mass balance of biomass is: $\mu X = F X$
Mass balance of methanol: $F M_{in} = F M_{out} + V q X$

The yield coefficient and the specific growth rate are then given by the following equations:

$$\mu = \frac{F}{V}$$

$$Y_{s/x} = \frac{q}{\mu} = \frac{(M_{in} - M_{out})}{X}$$

The results for the yield coefficient and the specific growth rate values are summarized in the Table below:

μ [1/h]	0.189	0.197	0.338	0.437	0.465
$Y_{s/x}$	2.2	2.13	1.85	1.75	1.76

EXAMPLE 6.5 (Section 6.2)

For the hydrolysis of soy protein, the endopeptidase from *Streptomyces griseus* was used. The reaction was performed in a CSTR hollow fiber system for continuous hydrolysis (Deslie, W.D., Cheyran, M. 1981. Biotechnol. Bioeng. **23**, 2257)[3].

The conversion ratio for the reaction has been defined as a ratio of nitrogen in the permeate to nitrogen in the feed. The concentration of the substrate in the feed was 1% (w). Flow rate of the substrate was $F=9$ ml/min. The volume of the reaction vessel was $V=550$ mL. The concentration of the enzyme used was $E=1$ g/L and the measured conversion band on the nitrogen balance was 90%. Evaluate the specific rate of hydrolysis.

Solution:

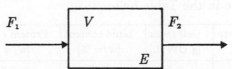

C...concentration of species based on nitrogen.

System: Content of tank and enzymatic reaction.

Because the system is filtered, the accumulation of compounds containing nonhydrolyzable nitrogen occurs. Observed conversion was time independent which means that the accumulation of hydrolyzable protein is zero and the system is operating at a quasi-steady state. The balances for the hydrolysis process can be written in the following form:

Overall balance:

$$INPUT \quad - \quad OUTPUT \quad = \quad 0$$
$$F_1 \quad - \quad F_2 \quad = \quad 0$$

Raw protein portion:

$$(Rate\ of\ input) - (Rate\ of\ hydrolysis) = \left(\begin{array}{c} Accumulation\ of \\ nonhydrolyzable\ protein \end{array} \right)$$

$$F_1 C_1 - V E r = \frac{\mathrm{d}(V C_N)}{\mathrm{d}t}$$

Soluble protein portion:

$$(Rate\ of\ hydrolysis) - (Rate\ of\ output) = 0$$
$$V\,E\,r - \quad F_2 C_2 = 0$$

Because conversion $y = C_1/C_2$ is 0.9, the balances with the reaction rate have to be rearranged in the following form:

$$r = \frac{F_1 C_1 y}{VE}$$

The resulting rate of protein hydrolysis is 0.00729 g protein/(g catalyst· min). The rate of accumulation of nonhydrolyzable protein is then 1.63×10^{-5} g(L of reactor volume· min)$^{-1}$.

EXAMPLE 6.6 (Section 6.2)

Microbial lipids (extracellular product) could be produced in a nitrogen limited continuous culture (Choi, S.Y. *et al.* 1982. Biotechnol. Bioeng. **24**, 1165)[1]. The experimental data from a cultivation of *Rhodotorula gracilis* (NRRL-Y- 1091) are in the Table below.

Dilution rate $D[h^{-1}]$	Cell mass [g DW/L]	Lipid content [w/w %]	Protein content [w/w %]
0.02	9.6	49.8	8.3
0.03	6.7	40.0	11.0
0.05	2.6	20.1	29.0
0.07	2.1	17.0	32.0
0.09	1.5	14.1	37.1

Evaluate the specific rates of lipid and protein biosynthesis.

Solution:

(The same mass balance principles and system flow chart will be used as in Examples 6.2 – 6.4. The mass balance for the intracellular product will be modified.)

Mass balance for the intracellular product in continuous-flow culture:

$$RATE\ OF\ BIOSYNTHESIS - RATE\ OF\ OUTPUT = 0$$
$$V\,r\,X \qquad - \qquad F\,X\,C \qquad = 0$$

Because $D = F/V$, the specific rate of biosynthesis of the intracellular product is given by the following equation:

$$r = D\,C$$

The summary of calculated results:

D [1/h]	0.02	0.03	0.05	0.07	0.09
$r_{prot} \times 10^3$ [1/h]	1.66	3.3	14.5	22.4	33.39
$r_{lipid} \times 10^3$ [1/h]	9.96	12	10.05	11.9	12.69

6.3 TRANSIENT MASS BALANCES

This section considers the formulation and solution of material and energy balances when input, output and accumulation vary with time. The analysis of a transient process yields a differential equation relating rate of input, rate of output and rate of accumulation. The only principles involved are the Conservation of Mass and the first law of Thermodynamics. The complete specification of the problem mathematically requires the statement of one or more initial conditions. Once the differential equation has been derived and the initial conditions formulated, the solution is a purely mathematical problem. The mathematical background required for this type of mass balances is the knowledge of how to solve first order ordinary differential equations.

Let us consider a system, such as schematically shown below, with one input and one output, the flows of both being functions of time.

$$\dot{m}_i \longrightarrow \left(SYSTEM \ \ M \right) \longrightarrow \dot{m}_o$$

$\dot{m}_i$ = mass rate of flow into the system, e.g. kg/s
$\dot{m}_o$ = mass rate of flow out of the system, e.g. kg/s
M = mass contained within the system, e.g. kg

Now choose a basis of Δt sec where Δt is a time short enough so that $\dot{m}_i$ and $\dot{m}_o$ may be considered constant. The overall material balance becomes

$$(INPUT) \ - \ (OUTPUT) \ = \ (ACCUMULATION)$$

$$\dot{m}_i(\Delta t) - \dot{m}_o(\Delta t) \ = \ M_{i+\Delta t} - M_t \qquad (6.9)$$

where the subscript t denotes the value at time t. Division by Δt yields

$$\dot{m}_i \;-\; \dot{m}_o \;=\; \frac{M_{t+\Delta t} - M_t}{\Delta t} \qquad (6.10)$$

We now let $\Delta t \rightarrow 0$ with the result that the right hand side becomes

$$\lim_{\Delta t \to 0} \frac{M_{t+\Delta t} - M_t}{\Delta t} \;=\; \frac{dM}{dt} \qquad (6.11)$$

Therefore, the instantaneous overall mass balance is

$$\dot{m}_i \;-\; \dot{m}_o \;=\; \frac{dM}{dt} \qquad (6.12)$$

which we may formulate in words as

$$\begin{bmatrix} RATE\ OF \\ INPUT \end{bmatrix} - \begin{bmatrix} RATE\ OF \\ OUTPUT \end{bmatrix} = \begin{bmatrix} RATE\ OF \\ ACCUMULATION \end{bmatrix}$$

This is the balance which applies at any instant of time for any number of input and output streams. When applied in this form, a basis need not be stated explicitly because the basis $\Delta t \rightarrow 0$ has already been used in deriving the equation. An analogous derivation may be made for a component as follows:

Let $\quad C_A \;=\;$ weight fraction of component A in output
$\quad\quad c_A \;=\;$ weight fraction of component A in input
$\quad\quad \bar{C}_A \;=\;$ average weight fraction of component A within the system

As before, for the basis Δt

$$(INPUT) \quad - \quad (OUTPUT) \quad = \quad (ACCUMULATION)$$

$$\dot{m}_i c_A(\Delta t) \quad - \quad \dot{m}_o C_A(\Delta t) \quad = \quad M\bar{C}_{A_{(t+\Delta t)}} - M\bar{C}_{A_t} \qquad (6.14)$$

Dividing by Δt and taking the limit, the component mass balance becomes

$$\dot{m}_i c_A \quad - \quad \dot{m}_o C_A \;=\; \frac{d(M\bar{C}_A)}{dt} \qquad (6.15)$$

Note that

$$\begin{bmatrix} RATE\ OF \\ ACCUMULATION\ OF \\ COMPONENT\ A \end{bmatrix} = \begin{bmatrix} THE\ DERIVATIVE \\ OF\ THE\ AMOUNT\ OF\ A \\ WITHIN\ THE\ SYSTEM \end{bmatrix}$$

the amount of A within the system is equal to the product of the total mass within the system and the average concentration of A within the system. A common error is to write the rate A accumulation as $M\frac{dC_A}{dt}$. This is incorrect unless the mass within the system is constant. The correct expression is given on the right hand side of Equation (6.15). The result of a transient mass balance is a differential equation rather than an algebraic equation. The solution of the differential equation requires much more information than the algebraic equation. We need flows and compositions as functions of time and information on conditions within the system which will enable us to calculate the average composition. Since the latter information is particularly difficult to obtain, frequently, idealizations of the system are used, called models.

EXAMPLE 6.7 (Section 6.3)

Water enters a tank at the rate of 25 L/min, it is being withdrawn at a rate which varies with time according to $25 \left(1 - e^{-0.1t}\right)$ L/min where t is in minutes. If the tank initially contains 50 L, how many gallons of water will the tank contain when steady state is reached?

Solution:

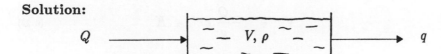

Q = volumetric feed rate, (L/min)
q = volumetric effluent rate, (L/min)
V = volume of water contained within the tank
ρ = density of water, (kg/L)

Looking at the form of the output rate we see that for very long times it approaches the input rate so that a state will be reached (asymptotically at $t \rightarrow \infty$) in which input and output are equal and thus volume of water within the tank will be constant.

System: contents of tank.

Since there is only one component, only an overall mass balance may be made.

$$\begin{bmatrix} RATE\ OF \\ INPUT \end{bmatrix} - \begin{bmatrix} RATE\ OF \\ OUTPUT \end{bmatrix} = \begin{bmatrix} RATE\ OF \\ ACCUMULATION \end{bmatrix} \tag{6.16}$$

$$Q\rho - q\rho = \frac{d(V\rho)}{dt} \tag{6.17}$$

and since the density, ρ, is constant

$$\frac{dV}{dt} + q = Q \tag{6.18}$$

if we now substitute for q:

$$q = Q(1 - e^{-at}) \tag{6.19}$$

where $Q = 25$ L/min and $a = 0.1$

$$\frac{dV}{dt} + Q(1 - e^{-at}) = Q \tag{6.20}$$

This is the differential equation to be solved with the initial condition at

$$t = 0 \qquad V = V_o \tag{6.21}$$

The solution is easily shown to be

$$V = -\frac{Q}{a} e^{-at} + K \tag{6.22}$$

Using the initial condition, Equation (6.12) becomes:

$$K = \frac{Q}{a} + V_o \tag{6.23}$$

and the complete solution is

$$V - V_o = \frac{Q}{a} (1 - e^{-at}) \tag{6.24}$$

As $t \to \infty$, V reaches a constant value, V_∞, given by

$$V_\infty = V_o + \frac{Q}{a} \qquad (6.25)$$

Substitution of the numerical values gives

$$V = 300 \text{ Liters}$$

There are several important conclusions from this Example (6.7) which should be noted:

Conclusions:

* If the densities of all streams are equal and constant, they cancel from the mass balance and we are left with a volume balance. The volume balance is only correct under these conditions.
* The equation resulting from each transient mass balance is a differential equation. For a solution, an initial condition is required.
* The number of independent transient material balances is equal to the number of species present. This is the same as for steady state and intermittent operation when no chemical reactions occur.

A problem will be considered where two species are present.

EXAMPLE 6.8 (SECTION 6.3)

A mixed tank contains 300 L of a 40% sugar solution. At time zero, a pure water flow of 2 L/min is started. The outlet stream from the tank is removed under level control so that the tank always contains the same volume of solution. For this system, the density is a function of composition. Over the range of interest, the density of the solution, ρ, is related to the weight fraction of sugar, x, as follows:

$$\rho = \rho_w(1 + ax) \qquad (6.26)$$

where

$$\rho_w = \text{density of water}$$
$$a = 0.98$$

How long will it take to reduce the sugar fraction in the exit stream to 10% of its initial value. What is the mass of solution in the tank at this time?

Solution:

let Q =   volumetric feed rate
q =   volumetric withdrawal rate
x =   sugar fraction
ρ =   density

System: the contents of the tank.

Since there are two species, we may write two material balances.

$$\begin{bmatrix} RATE\ OF \\ INPUT \end{bmatrix} - \begin{bmatrix} RATE\ OF \\ OUTPUT \end{bmatrix} = \begin{bmatrix} RATE\ OF \\ ACCUMULATION \end{bmatrix}$$

Overall balance: $\qquad Q\,\rho_w - q\,\rho = \dfrac{dV\bar{\rho}}{dt}$ (6.27)

Sugar balance: $\qquad -q\,\rho\,x = \dfrac{dV\bar{\rho}x}{dt}$ (6.28)

where the overscores denote average values in the tank.

Since V is constant with level control,

$$Q\,\rho_w - q\,\rho = V\frac{d\bar{\rho}}{dt} \qquad (6.29)$$

$$-q\,\rho\,x = V\frac{d\bar{\rho}x}{dt} \qquad (6.30)$$

At this point, information is needed on the average density and concentration in the tank. Since ρ and x vary with position in the tank and this variation is seldom known, the following idealization is made.

It is assumed that the contents of the tank are mixed so well that there is no spatial variation of any property within the tank. Since the outlet stream comes from the tank, the properties (density, concentration, etc.) of the outlet stream are the same as those inside the tank. The equipment operating under this assumption is called the *well mixed tank, perfectly mixed tank* or *well stirred tank*.

6.3.1 A Perfectly Stirred Tank Model

This model of an operation is based on theoretical assumptions of perfect and instantaneous mixing in the tank which results in a perfectly homogeneous tank contents. Note that the stream entering the tank does not have the same composition as the tank contents. A tank containing a low viscosity liquid agitated with a stirrer approximates this behaviour. However, it should be kept in mind that the well mixed tank is a model of reality and real tanks do not behave precisely as the model. Nevertheless, it is a useful model which gives a limiting behaviour for real systems.

Assuming a mixed vessel operating according to the assumption above as a perfectly stirred tank, we can write that

$$\bar{\rho} = \rho \quad ; \quad \bar{x} = x$$

The differential equations (6.29) and (6.30) then become respectively

$$Q \, \rho_w - q \, \rho = V \frac{d\rho}{dt} \tag{6.31}$$

$$-q \, \rho \, x = V \, x + V \rho \frac{dx}{dt} \tag{6.32}$$

These equations must be solved subject to the initial conditions at

$$t = 0 \qquad\qquad \rho = \rho_o \qquad\qquad x = x_o \tag{6.33}$$

Substituting Equation (6.31) into Equation (6.32) we obtain

$$\frac{\rho}{\rho_w} \frac{dx}{dt} + \frac{Q}{V} x = 0 \tag{6.34}$$

Substitution of the density relation, Equation (6.26), gives

$$\frac{1}{x} + a \frac{dx}{dt} = \frac{Q}{V} \tag{6.35}$$

which can be integrated to give

$$x + \ln x = -\frac{Q}{V} t + K \qquad (6.36)$$

From the initial condition

$$K = ax_o + \ln x_o \qquad (6.37)$$

The complete solution is

$$\frac{x}{x_o} e^{a(x-x_o)} = \exp\left(-\frac{Q}{V}t\right) \qquad (6.38)$$

Substitution of the numerical values from EXAMPLE 6.7 gives time

$$t = 398 \text{ min}$$

for the concentration to drop to 10% of its initial value. The mass in the tank at any time is found simply from the product of the volume and density.

Initial mass = $300 [1 + 0.98(0.4)]$ = 417.6 kg

Final mass   = $300 [1 + 0.98(0.04)]$ = 301.7 kg

From the results of EXAMPLE 6.8 the following points should be noted:

Conclusions

* Input and output volumetric flows are not equal when the density is variable.
* Solution of transient problems is facilitated by working with symbols as long as possible and substituting numerical values after integration of the differential equations.

6.3.2 Transient Mass Balances With Reaction

The treatment of transient balances for systems with reaction is identical to that for systems without reaction if the mass balances are made on components (as defined previously). Therefore, for a component, Equations (6.13) and (6.16) are correct. However, if a reaction is occurring there is frequently additional information available in the form of experimental data or a correlating equation relating the rate of reaction to temperature, pressure and composition. For the moment, it could be assumed that we have information on the rate of a chemical reaction in the system sketched below.

let Q = volumetric rate of input
q = volumetric rate of output
C_i = volumetric concentration of i in the inlet stream
c_i = volumetric concentration of i in the outlet stream
$\bar{c}_i$ = average volumetric concentration of i in the system
V = volume of the system

Assume that the following reaction occurs: $A + 2B \rightarrow C + 3D$ and that it occurs at an average rate within the system of $\bar{r}\ V$ where
$\bar{r}$ = average molar rate of disappearance of A within the system per unit volume, moles/(volume)(time)

Thus, the quantity $\bar{r}\ V$ is the rate of disappearance of A within the system. The quantity r is positive. From the stoichiometry of the reaction we see that $\bar{r}$ can also be given as

$\bar{r}$ = average rate of appearance of C within the system per unit volume
or $\bar{r}$ = one-half of the average rate of disappearance of B within the system per unit volume
or $\bar{r}$ = one-third of the average rate of appearance of D within the system per unit volume.

We now write equations which account for the fate of each species. The general form is

$$\begin{bmatrix} RATE\ OF \\ INPUT \\ OF\ SPECIES\ i \end{bmatrix} - \begin{bmatrix} RATE\ OF \\ OUTPUT \\ OF\ SPECIES\ i \end{bmatrix} = \begin{bmatrix} RATE\ OF \\ ACCUMULATION \\ OF\ SPECIES\ i \end{bmatrix} \qquad (6.39)$$

Since we are following species we must account for the fact that they can appear and/or disappear through reaction. Therefore, the rates of input and output are defined as follows:

$$
\begin{bmatrix} RATE\ OF \\ INPUT \\ OF\ SPECIES\ i \end{bmatrix} = \begin{bmatrix} RATE\ AT\ WHICH \\ i\ ENTERS \\ THE\ SYSTEM \end{bmatrix} + \begin{bmatrix} RATE\ AT\ WHICH \\ i\ IS\ FORMED \\ BY\ REACTION \end{bmatrix} \tag{6.40}
$$

$$
\begin{bmatrix} RATE\ OF \\ OUTPUT \\ OF\ SPECIES\ i \end{bmatrix} = \begin{bmatrix} RATE\ AT\ WHICH \\ i\ LEAVES \\ THE\ SYSTEM \end{bmatrix} + \begin{bmatrix} RATE\ AT\ WHICH \\ i\ IS\ CONSUMED \\ BY\ REACTION \end{bmatrix} \tag{6.41}
$$

The definition of the accumulation is the same as that given in Equation (6.16). Returning now to the system shown in the Figure above, we write the four species balances in the molar form:

$$
A: \qquad QC_A - (qC_A + \bar{r}V) = \frac{d(V c_A)}{dt} \tag{6.42}
$$

$$
B: \qquad QC_A - (qC_A + 2\bar{r}V) = \frac{d(V c_A)}{dt} \tag{6.43}
$$

$$
C: \qquad (QC_A + \bar{r}V) - qC_C = \frac{d(V c_C)}{dt} \tag{6.44}
$$

$$
D: \qquad (QC_A + 3\bar{r}V) - qC_D = \frac{d(V c_D)}{dt} \tag{6.45}
$$

We have taken C_i and c_i to be molar concentrations and $\bar{r}$ to be the average molar rate of reaction per unit volume.

EXAMPLE 6.9 (Section 6.3.2)

A well-stirred tank is fed with N (gmoles/min) of pure CO and M (gmoles /min) of pure O_2. The molar density of the gases within the tank is ρ (gmoles/cm^3). The tank volume is C (cm^3). Within the tank the following reaction occurs:

$$CO \quad + \quad O_2 \quad \longrightarrow \quad CO_2$$

at a rate of r (moles) of CO consumed per cm^3 of volume per min. Write the carbon and oxygen balances (component balances) and the species balances.

Solution:

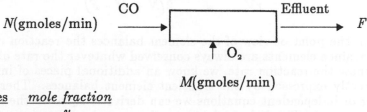

species	mole fraction
O_2	y
CO_2	x
CO	$1-x-y$

System: the tank.

We take the effluent flow as F gmoles/min and assign mole fractions as indicated in the sketch. For this system there are two components (three species and one reaction). We write the element or component balances noting that elements are neither created nor destroyed.

C-balance $\qquad N \;-\; F(1-y) \;=\; \dfrac{d[(1-y)V\rho]}{dt}$ (6.46)

O-balance $\quad N + 2M \;-\; F(1+x+y) \;=\; \dfrac{d[(1+x+y)V\rho]}{dt}$ (6.47)

We note that since elements are conserved, the reaction rate does not appear in these equations. We now write the three species balances noting that the rate of formation of CO_2 is equal to the rate of consumption of CO and that the rate of consumption of O_2 is double the rate of consumption of CO.

CO-balance $\qquad N \;-\; F(1-x-y) \;-\; rV \;=\; \dfrac{d(1+x+y)V\rho}{dt}$ (6.48)

CO_2-balance $\qquad\qquad rV \;-\; Fx \;=\; \dfrac{d(xV)}{dt}$ (6.49)

O_2-balance $\qquad\qquad M \;-\; Fy \;-\; \frac{rV}{2} \;=\; \dfrac{d(xV\rho)}{dt}$ (6.50)

We now compare the two sets of equations. it is obvious that if equation (6.48) is added to Equation (6.49) the result is identical to Equation (6.46).

Similarly, if (6.48) is added to two times (6.49) and to two times (6.50), the result is identical to (6.47). Therefore, the species balances, Equations (6.48) – (6.50), are consistent with the element balances. It is important to note that Equations (6.48), (6.49) and (6.50) are independent, i.e., the number of independent equations is equal to the number of species in this case rather than the number of components as we found in the steady state case. The additional independent equation appears because we have a new quantity here: r, the rate of the reaction. This new quantity is related to flows in the system. This is easily seen by solving Equations (6.46) and (6.50) to give

$$rV \quad = \quad 2(N + M - F) \tag{6.51}$$

From the point of view of the element balances the reaction rate is immaterial since elements are always conserved whatever the rate of reaction. If we know the reaction rate, we know an additional pieces of information not directly expressed by the transient element balances. Therefore, the number of independent equations we can derive is equal to the number of species present. The independent equations can be taken in a number of ways. For example, we could take Equations (6.46) and (6.47) as element balances and (6.49) as the definition of the reaction rate. Alternatively we could take Equations (6.46), (6.47) and any one of (6.48), (6.49) or (6.50). Or still another alternative would be Equations (6.48), (6.49) and (6.50).

We will now look at several specific examples.

EXAMPLE 6.10 (Section 6.3.2)

For the situation described in EXAMPLE 6.9 consider the following additional conditions:

a)   the tank volume V is constant
b)   the molar density of the gases within the tank, ρ, is constant. This corresponds to constant temperature and pressure if the gas mixture is ideal
c)   the rate of reaction r is constant
d)   the tank initially contains pure CO_2

Find the variation of the mole fractions with time.

Solution:

The process flowchart from EXAMPLE 6.9 depicts the system. Using (a) and (b) above, Equations (6.49) and (6.50) simplify to

$$r\,V - F\,x = V\,\rho\,\frac{dx}{dt} \tag{6.52}$$

$$M - F y - \frac{rV}{2} = V \rho \frac{dV}{dt} \tag{6.53}$$

The initial conditions are obtained from (d) above

$$\text{at} \qquad t = 0 \qquad x = 1 \tag{6.54}$$

$$\text{at} \qquad t = 0 \qquad y = 0 \tag{6.55}$$

Since r is constant, Equations (6.52) and (6.53) can be solved to give

$$x = K_1 e^{-Ft/V\rho} + \frac{rV}{F} \tag{6.56}$$

$$y = K_2 e^{-Ft/V\rho} + \frac{M}{F} - \frac{rV}{2F} \tag{6.57}$$

From Equations (6.54) and (6.55) we obtain

$$K_1 = 1 - \frac{rV}{F} \tag{6.58}$$

$$K_2 = \frac{rV}{2F} - \frac{M}{F} \tag{6.59}$$

Thus the solution for x and y is

$$x = e^{-Ft/V} + \frac{rV}{F}(1 - e^{-Ft/V\rho}) \tag{6.60}$$

$$y = (\frac{M}{F} - \frac{rV}{2F}) (1 - e^{-Ft/V\rho}) \tag{6.61}$$

We can also eliminate F from Equations (6.60) and (6.61) through Equation (6.51):

$$F = N + M - \frac{rV}{2} \tag{6.62}$$

to give x and y in terms of input streams and system properties. The CO mole fraction, z, is easily obtained from

$$z = 1 - x - y \tag{6.63}$$

It is interesting to look at the mole fractions for long times, $t \to \infty$:

$$CO_2 : \qquad x_\infty = \frac{rV}{F} \tag{6.64}$$

$$O_2 : \qquad y_\infty = 1 - \frac{N}{F} \tag{6.65}$$

$$CO : \qquad z_\infty = \frac{N}{F} - \frac{rV}{F} \tag{6.66}$$

Eliminating F through Equation (6.62):

$$CO_2 : \qquad x_\infty = \frac{rV/N}{1 + \frac{M}{N} - \frac{rV}{2N}} \tag{6.64}$$

$$O_2 : \qquad y_\infty = 1 - \frac{x_\infty}{rV/N} \tag{6.65}$$

$$CO : \qquad z_\infty = x_\infty \left(1 - \frac{1}{rV/N}\right) \tag{6.66}$$

Since $0 < (x_\infty, y_\infty, z_\infty) < 1$ there are limits on the values of rV/N and M/N which are necessary to have all three species in the exiting stream. For example if $rV/N > 1.0$, then there is no CO $(z_\infty = 0)$ and for x_∞ and y_∞ to be properly bounded $M/N > 0.5$. In transient problems as in steady state problems we can check the reasonableness of the answers. One simple check is to look at the results of the transient analysis for $t \to \infty$.

The ease with which the differential equations can be solved analytically depends upon the form of the reaction rate expression. We now look at a first order rate expression used in the following EXAMPLE (6.11).

EXAMPLE 6.11 (Section 6.3.2)

A holding tank is required to reduce the concentration of an isotope in water. The water flow rate is 10 L/min and the isotope concentration is 0.001 mg/L. How big should the tank be to reduce the isotope concentration by a factor of ten at steady state operating conditions? If the tank is initially filled with pure water when the contaminated water first enters, how does the effluent concentration change with time? The isotope half-life is 10 minutes and the decay reaction is first order.

Solution:

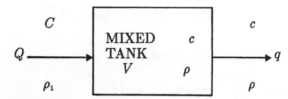

System: the contents of the tank

let Q, q = volumetric flow rates (L/min)
C, c = volumetric concentrations of isotope (mg/L)
ρ_1 = density of feed (mg/L)
ρ = density of effluent (mg/L)
V = tank volume (L)

The reaction for the isotope may be written: A $\rightarrow$ Product
For a first order reaction the half life, $t_{1/2}$, is related to the reaction rate by

$$-\frac{dc}{dt} = \frac{0.693}{t_{1/2}} C = r \left(\frac{mg}{L\ min}\right) \tag{6.67}$$

where r is the rate of disappearance of the isotope per unit volume $(-dc/dt)$.
We assume perfect mixing and write the two independent material balances. First we examine the water balance.

Water: $$Q(\rho_1 - C) - q(\rho - c) = \frac{d[V(\rho - c)]}{dt} \tag{6.68}$$

$$Q(\rho_1 - C) - q(\rho - c) = V \frac{d(\rho - c)}{dt} \tag{6.69}$$

Since $(\rho_1, \rho) >> (C, c)$; $\rho_1 \sim \rho$, the water balance yields:

$$Q = q \tag{6.70}$$

Since the solution is so dilute we could have stated this already on the basis of EXAMPLE 6.7, Equation (6.18).

We now look at the isotope balance in this example.

$$\text{Isotope:} \qquad QC - qc - Vr = \frac{d(Vc)}{dt} \qquad (6.71)$$

Substitution of previous results yields

$$QC - Qc - Vkc = V\frac{dc}{dt} \qquad (6.72)$$

where
$$k = 0.693/t_{1/2} \qquad (6.73)$$

At steady state
$$\frac{dc}{dt} = 0 \qquad \text{and}$$

$$V = \frac{Q(C-c)}{kc} \qquad (6.74)$$

$$V = \frac{Q}{k}\left(\frac{C}{c} - 1\right) \qquad (6.75)$$

$$V = \frac{Q}{0.693} t_{1/2}\left(\frac{C}{c} - 1\right) \qquad (6.76)$$

Substitution of numerical values yields

$$V = 1,298 \text{ L}$$

The time variation of concentration is found by solving

$$\frac{dc}{dt} + \left(k + \frac{Q}{C}\right)c = \frac{Q}{V}C \qquad (6.77)$$

subject to the initial condition

$$\text{at} \quad t = 0, \qquad c = 0 \qquad (6.78)$$

The solution to this first order equation is readily found to be

$$\frac{dc}{dt} = K e^{-(k+\frac{Q}{V})} + \frac{Q/V}{k + Q/V} C \qquad (6.79)$$

From the initial condition

$$K = -\frac{Q/V}{k + Q/V} C \qquad (6.80)$$

Therefore,
$$\frac{c}{C} = \frac{Q/V}{k + Q/V} [1 - \exp\{-(k + \frac{Q}{V}) t\}] \qquad (6.81)$$

Note that in the system without reaction, the outlet stream concentration changes in an exponential fashion from zero to its steady state value as a response to the inlet stream concentration change.

We now look at a situation where a dilute solution reacts at a rate r.

EXAMPLE 6.12 (Section 6.3.2)

A well mixed tank similar to that of EXAMPLE 6.11 is required for a reaction which occurs consuming species A at a constant rate of r kg/L·h. If the concentration of reacting material in the tank is initially zero, and at time zero a dilute solution of reactant flows into the tank, what is the time variation of the concentration in the stream leaving the tank? Assume:

1. the reaction is A → products
2. the entering stream is a dilute solution of A
3. densities are constant
4. the tank volume is constant
5. the tank is well mixed.

Solution:

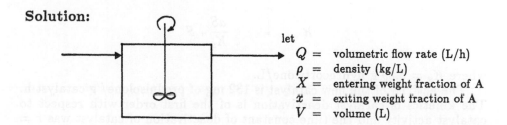

let

Q = volumetric flow rate (L/h)
ρ = density (kg/L)
X = entering weight fraction of A
x = exiting weight fraction of A
V = volume (L)

System: the contents of the tank

Since the entering solution is dilute, the densities are constant and the volume contained in the tank is constant, the entering and exiting volumetric flow rates are equal. No further information is obtained from the total mass balance.

We now make a balance on species A noting that the tank is well mixed.

$$ Q\,\rho\,X \quad - \quad Q\,\rho\,x \quad - \quad r\,V \;=\; \frac{\mathrm{d}(V\,\rho x)}{\mathrm{d}t} \tag{6.82} $$

Using the constancy of V and ρ:

$$ \frac{V}{Q}\frac{\mathrm{d}x}{\mathrm{d}t} \;+\; x \;-\; X \;=\; -\frac{r}{\rho Q} \tag{6.83} $$

The initial condition is

$$ \text{at} \quad t = 0\,, \quad x = 0 \tag{6.84} $$

The solution of the differential equation yields

$$ \frac{x}{X} \;=\; (1 \,-\, \frac{r}{\rho QX})\,(1 \,-\, e^{-Qt/V}) \tag{6.85} $$

EXAMPLE 6.13 (Section 6.3.2)

Microbial transformation of steroid hydrocortisone into prednisolone may be performed using immobilized *Corynebacterium simplex* cells. (Constantinides, A. 1980. Biotechnol. Bioeng. **22**, 119-186)[2]. The rate of reaction kinetics can be described by the Michaelis–Menten equation:

$$ R \quad = \quad \frac{aS}{K_s} + S $$

where $K_s = 5.39$ g hydrocortisone/L.

The activity of the new catalyst is 132 mg of prednisolone/ g catalyst·h. The kinetics of catalyst deactivation is of the first order with respect to catalyst activity and the time constant of deactivation of catalyst was $\tau = 105$ h.

When 0.25 kg of biocatalyst is used periodically in a 12 L laboratory tank for treatment of 10 L of substrate containing 50 g/L of hydrocortisone, how long would be the operation cycle after 0, 2.5 and 5 days of catalyst operation if we desire 0.98 weight yield of prednisolone from hydrocortisone (the difference in molecular weights of the substrate and the product can be neglected)?

Solution:

System: batch reactor.

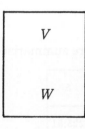

Nomenclature:
W ... catalyst weight
V ... volume of substrate
a ... catalyst activity
T ... operation time of catalyst
before reaction

Two balances for the batch reactor have to be written. The first one for the bioconversion and the second one for the catalyst deactivation.

Balance for the prednisolone production:

$$RATE\ OF\ CONVERSION \quad = \quad RATE\ OF\ ACCUMULATION$$

$$W\ \frac{a(T+t)S}{S+K_s} \quad = \quad \frac{d(VP)}{dt}$$

Balance for the catalyst activity:

$$RATE\ OF\ DEACTIVATION \quad = \quad RATE\ OF\ ACCUMULATION$$

$$-\frac{1}{\tau}\ a\ W \quad = \quad \frac{d(Wa)}{dt}$$

Initial conditions:
at $t = T$: $c = 50$ g hydrocortisone/L
at $t = 0$: $a = 132$ g prednisolone/(kg catalyst·h)

Method of solution: The balance of catalyst activity can be solved analytically.

$$a(T+t) \ = \ 132\ \exp[-(T+t)/\ \tau]$$

For the solution of the product balance it is more comfortable to use a simple computer program for numerical integration (for example the explicit Euler method with steps in time of 0.01h). The program was written in BASIC for the IBM AT Personal Computer and its listing is as follows:

```
10 REM steroid transformation
20 INPUT''initial time (day)'';T
30 T=T*24!
40 P=0
50 P=P+.24.025*5.5*EXP(-T/105!)*(50!-P)/(50!-P+5.39)
60 T=T+.01
70 IF P/50! > .98 THEN GOTO 90
80 GOTO 50
90 PRINT''operation cycle is'';T;''hrs''
100 STOP
```

The results for the length of the operation cycle are summarized below:

start [day]	0	2.5	5
operation cycle [h]	23.72	106.58	225.51

From the solutions, and particularly those to EXAMPLES 6.11 and 6.12, the following conclusions can be drawn concerning operating characteristics of a perfectly stirred tank reactor:

Conclusions

* For dilute solutions the input and output volumetric flow rates are (essentially) equal. This fact is obtained from the solvent balance, therefore, the number of independent balances remaining (under the assumption $Q = q$) is one less than the number of species.
* Feed and effluent concentrations will not be equal at steady state if a reaction occurs in the system, i.e. species will not be conserved.
* when there is no reaction ($r = 0$) and the inlet concentration changes suddenly, the outlet concentration responds exponentially. For $t \to \infty$, the inlet and outlet concentrations are equal.
* when there is reaction and $t \to \infty$, the inlet and outlet concentrations are not equal.

6.3.3 The Plug–Flow Model

In connection with EXAMPLE 6.11 we might look at another model for the system. This model, called plug flow or piston flow, is an idealization which assumes that no mixing occurs within the system. It is as if the system were a long pipe and at time zero an impermeable membrane separated the incoming fluid from that already in the system. We might visualize this as follows:

The time required for the membrane to move down the pipe is simply V/Q. The outlet concentration would be zero until $t = V/Q$ at which time it would jump to its steady state value. If we look at a small piece of the contaminated fluid of Example 6.10 as it flows from one end of the system to the other we see that it must remain in the system long enough for the concentration to fall to 10% of its initial value.

This time found from the differential equation

$$-\frac{dc}{dt} = k\,c \qquad (6.86)$$

subject to the initial conditions:

$$\text{at} \quad t = 0 \qquad c = C \qquad (6.87)$$

The solution is $\qquad \ln c = -kt + K \qquad (6.88)$

and after using the initial condition

$$t = \frac{1}{k} \ln \frac{C}{c} = \frac{t_{1/2}}{0.693} \ln \frac{C}{c} \qquad (6.89)$$

Substitution of numerical values from Example 6.11 yields

$$t = 33.22 \text{ min}$$

Now we compute the volume $\qquad V = t\,Q$

And according to values in Example 6.11 the volume is

$$V = 332.2 \text{ L}$$

Using the plug flow the required volume is much smaller.

6.4 SUMMARY OF THE REACTOR TYPES

In terms of physical configurations encountered, there are basically only two types of reactors: the tank and the tube.

6.4.1. The Continuous Stirred Tank Reactor (CSTR)

This type of reactor consists of a well-stirred tank into which there is a continuous flow of reacting material, and from which the (partially) reacted material passes continuously. It is as wide as they are deep) that good stirring of their content is essential, otherwise there could occur a bulk streaming of the fluid between inlet and outlet and much of the volume of the vessels would be essentially dead space.

The important characteristic of a CSTR is the stirring. The most appropriate first approximation to an estimation of its performance is based on the assumption that its contents are perfectly mixed. As a consequence, the effluent stream has the same composition as the contents and this demonstrates the important distinction between the CSTR and the tubular reactor. A fair approximation to perfect mixing is not difficult to attain in a CSTR, provided that the fluid is not too viscous.

6.4.2. The Tubular Reactor

The tubular reactor is so named because in many of its instances it takes the form of a tube. However, what is meant in general by a tubular reactor is any continuously operating reactor in which there is a steady state movement of one or all of the reagents in a chosen spatial direction (the reagents entering at one end of the system and products leaving at the other) and in which no attempt is made to induce mixing between elements of fluid at different points along the direction of flow: each element is assumed to move in a "plug flow" manner. Theoretically, the velocity profile of the fluid at a given cross section is assumed to be flat and it is also assumed that there is no axial diffusion and back-mixing of fluid elements.

Tubular reactors are also used extensively for catalytic reactions. Here the reactor is packed with particles of the solid catalyst and for this reason is often referred to as a **Fixed Packed-Bed Reactor (FPBR)**.

6.4.3 Other types of (bio)reactors:

a) Fluidized–bed reactor:
In this type of reactor the solid material in the form of fine particles is contained in a vertical cylindrical vessel. The fluid stream is passed upwards through the particles at a rate great enough for them to lift, but no so great that they are prevented from falling back into the fluidized phase above its

free surface by carry-over in the fluid stream. The bed of particles in this condition presents the appearance of boiling.

b) Bubble–phase reactor

Here a reagent is bubbled through a liquid with which it can react, because the liquid contains either a dissolved involatile catalyst or another reagent. The product may even be removed from the reactor in the gas stream.

c) Slurry–phase reactor

This type of reactor is similar to bubble-phase reactor, but the "liquid" phase consists of a slurry of liquid and fine solid catalyst (microbial cells or immobilized enzymes) particles.

d) Trickle–bed reactor

In this reactor, the solid catalyst is present, not as fine fluidized particles but as a fixed bed. The reagent which may be two partially miscible fluids, are passed either co-currently or counter-currently through the bed.

e) Moving–burden bed reactor

Here a fluid phase passes upwards through a packed bed of solid. Solid is fed to the top of the bed, moves down the column in a closely plug-flow manner, and is removed from the bottom.

6.4.4 Comparison of the Plug Flow and Well Mixed Tank Models

The transient behaviors of the well mixed system and the plug flow system are quite different. Their transient response curves for the situation where a step function was introduced at the inlet point (e.g. continuous flushing of the contents with water) are sketched below.

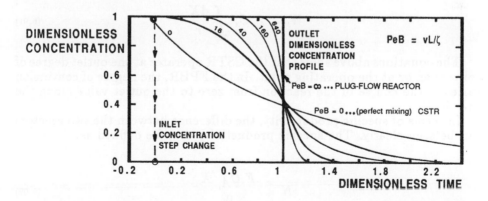

Figure 6.1: Response of continuous-flow bioreactors to the inlet concentration step change

The well mixed tank reaches steady state asymptotically while the plug flow system jumps from the initial to the final state. These curves emphasize again the fact that the transient behaviour of a system depends on factors within the system. In steady state operations, on the other hand, we do not need explicit information on the interior workings of the system.

The most common reactors used in bioprocesses is the batch mixed reactor, CSTR and FPBR. In the CSTR, the catalyst or biomass is suspended in the liquid with or without immobilization. In the FPBR, the catalyst or biomass are immobilized into desirable particulate or granular form and packed in the reactor.

6.4.5 CSTR or FPBR ?

In terms of a simple reactor mass balance which serves as a basis for expressing the specific productivity, the CSTR and the FPBR could be compared as follows[8]:

The design equation of CSTR:

$$F\left(C_{A_i} - C_{A_e}\right) = r\,W \tag{6.90}$$

The design equation of FPBR:

$$r = \frac{dC_A}{d(W/F)}$$

by integrating, gives:

$$W = F\,C_A \int \frac{dX}{r} \tag{6.91}$$

The equations above show that the CSTR operates at the outlet degree of conversion or at the operating point. In the FPBR, the degree of conversion varies in the course of the reaction from zero to the outlet value along the bed depth.

In terms of specific productivity, the difference between the two reactors can be seen clearly. The specific productivity may be defined as:

$$N = \frac{F\,C_{A_i}\,X}{W} \tag{6.92}$$

where:

F = feed rate

C_A = concentration of compound A; subscripts i and e represent influent and effluent, respectively

X = degree of conversion

r = rate of reaction

W = weight of catalyst

N = specific productivity (productivity per weight of catalyst)

Substitution of Equation (6.90) and Equation (6.91) into Equation (6.92) gives

$$N_{CSTR} = \frac{F \, C_{A_i} X}{W} = r \qquad (6.93)$$

$$N_{FPBR} = \frac{X}{\int \frac{dX}{r}} \qquad (6.94)$$

As an illustration, suppose we have a reaction:

$$A \longrightarrow P$$

$$-r_A = k \, C_A^{\,2} = k \, C_{A_i}^{\,2} \, (1 - X)^2 \qquad (6.95)$$

Substitution of Equation (6.95) into Equation (6.93) and Equation (6.95), respectively, yields:

$$N_{CSTR} = k \, C_{A_i}^{\,2} \, (1 - X)^2 \qquad (6.96)$$

$$N_{FPBR} = \frac{X}{\int \frac{dX}{kC_{A_i}^{\,2}(1 - X)^2}} = k \, C_{A_i}^{\,2} X \, (1 - X) \qquad (6.97)$$

For simplicity, let $k = 1$, $C_{A_i} = 1$. Equation (6.96) and Equation (6.96) become:

$$N_{CSTR} = (1 - X)^2 \qquad (6.98)$$
$$N_{FPBR} = X \, (1 - X) \qquad (6.99)$$

For $X = 0.1$ to $X = 0.9$

X	N_{CSTR}	N_{FPBR}
0.1	0.81	0.09
0.2	0.64	0.16
0.3	0.49	0.21
0.4	0.36	0.24
0.5	0.25	0.25
0.6	0.16	0.24
0.7	0.09	0.21
0.8	0.04	0.16
0.9	0.01	0.09

From this illustration, it is clear that at a high degree of conversion (X=0.9), the specific productivity of the FPBR is nine times as high as that of the CSTR. In fact, the actual relationship depends on the type of reaction rate. For the reaction of a positive order (the reaction rate decreases with the degree of conversion), the CSTR has a lower specific productivity than the FPBR. This drawback is more serious with increasing reaction order and with increasing outlet degree of conversion.

However, in a further consideration, the FPBR has another limitation. The diameter of catalyst should be around 2-10 mm [8]. The small size of the pellet in the FPBR leads to increased hydraulic resistance of the particle bed which is associated with higher operating cost for pumping. On the other side, the bigger size of the pellet causes the catalyst to become less effective because of an internal diffusion effect.

Moreover, the mechanical strength of the pellet must be sufficient enough to withstand the pressure without breaking and disintegrating.

The effect of internal diffusion could be accounted for quantitatively by introducing the effectiveness factor:

$$\eta = \frac{r}{r_o} \tag{6.100}$$

where:

r = the reaction rate affected by internal diffusion

r_o = the reaction rate under the same conditions when intraparticle diffusion has been eliminated, i.e. on a very small grain of catalyst.

When the catalyst pellet is very finely powderized, the effectiveness factor approaches unity, meaning that the active surface is fully utilized. In the CSTR where the catalysts are suspended in the liquid phase, a very small size of the catalyst can be adopted but for the reason of excessive pressure drop limitation it could not be used in the FPBR.

One advantage of the CSTR, apart from its simplicity of construction, is the ease of pH and temperature control facilitated by good mixing. Additional submerged cooling coil can be provided. The internal surface of the CSTR is also easier to clean and maintain deposit free.

6.4.6 The Catalyst Retention Reactor

This type of reactor which is a modification of the CSTR offers some advantages over the conventional CSTR. The individual, relatively small, particles of catalyst or biomass are kept free in the suspended state providing for unimpeded mass transfer and large surface area contact between the reactants and the catalyst. While the liquid is flowing through the reactor system on a continuous flow-through basis, the catalysts are retained within the reactor in a suspended state. This mode of operation can be labelled as a "hydraulic immobilization" of the solid catalyst phase. The tangential cross-flow solid-liquid separation system built into this reactor is self cleaning. When the filter element rotates, the "slip-flow" and the centrifugal forces at its surface prevent deposition of particles and surface clogging. This *in situ* solid-liquid separation requires less energy for rotation when compared to external centrifugation. It also reduces operating problems such as increased culture contamination or disturbance probabilities.

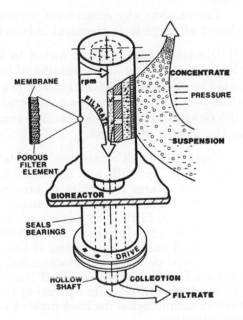

Figure 6.2:

A schematic cut-off diagram
of the rotating, tangential-flow,
self-cleaning filter element used
in the Catalyst Retention Reactor.

6.5 REACTORS IN THE PROCESS

6.5.1 Mode of Reactor Operation

Reactors may be operated in a variety of modes: batch, semi-batch, and continuous flow.

In batchwise operation the change of composition occurs in the time co-ordinate. Whether or not a batch system is uniform throughout its spatial co-ordinates. it always changes from moment to moment, and does so far as along as is needed for thermodynamic equilibrium to be attained (or until the process is brought to an end).

Alternatively, the reactants are continuously supplied into the reactor and at the same time an equal volume of reactor contents is discharged in order to maintain a constant level in the reactor. This mode of operation is referred to as a continuous mode of operation.

Certain processes are neither unambiguously batch nor unambiguously continuous, but should be described as "semi-batch" or "semi-continuous". For example, penicillin is made in a fermenter which is inoculated with the penicillin-producing microorganism at the start of a production run. After many hours, the contents of the fermenter are emptied, and penicillin removed from them. This would, therefore, seem to be a batch process. However, during the run, air and nutrients such as sugar, are continuously added to the fermenter, and gaseous waste products are continuously removed.

The reasons why continuous processes have been eventually adopted in almost all large-scale chemical industries are mainly as follows:

(i) Diminished labor costs, owing to the elimination of operation such as the repeated filling and emptying of batch vessels.

(ii) The facilitation of automatic control. This also reduces labor cost, although it usually requires considerable capital outlay.

(iii) Greater constancy in reaction conditions and hence greater constancy in the quality of product.

Certain chemicals produced in relatively small quantities such as pharmaceuticals, dyestuffs and so on are made batchwise. In a typical factory which has a large repertoire of products, each produced on a fairly small scale, batchwise processes are usually adopted. Such a system gives great flexibility. A further advantage of batchwise operations is that the capital cost is often less than for a corresponding continuous process when the desired rate of production is low. The possible long-term instability of microbial cultures and biochemical catalysts also mitigates against continuous flow bio-processes. Traditions and the training and background of the workforce involved in operating the bioprocess plays an important role even in technological decision making concerning the process and bioreactor configuration.

6.5.2 The Nature Of The Reactor Design Problem

The primary problems facing the engineer in reactor design are: how to choose the best type of a reactor for a particular chemical reaction; how to estimate its necessary size and how to determine its best operating conditions. The design of an industrial reactor must encompass an analysis of the whole technology. The engineer is required to decide which of many design alternatives is the most favorable. There are two usually relatively fixed factors which must be considered for the reactor design: the scale of operation (i.e. the required daily product output) and the kinetics of the given reaction. The overall plant design and/or the biocatalyst stability may dictate the choice of either a batch process or one of the several different kinds of continuous processes. There may be some constraints on the initial concentration of the reagents and also on the operating temperature and pressure. Controlled alterations in some of these variables can be made even during the course of the reaction.

Usually there are many combinations of operating conditions and the reactor size and/or type that will meet the requirements imposed by nature in terms of the reaction rate expression involved and those imposed by management in terms of required production capacity. The plant will require an overall optimization as a whole. The considerations associated with upstream and downstream processing may be eventually the deciding factor in chosing the reactor type. Simplicity in the plant and bioreactor design may ultimately represent a real economic benefit[4].

6.5.3 Conclusion

- There is no generally applicable method for selecting reactor types.
- CSTR gives flexibility and easiness which lead to cost reduction.
- Despite its other limitations, the FPBR gives higher specific productivity than CSTR.
- The Catalyst Retention Reactor offers advantages which overcome the shortcomings of the conventional CSTR.
- The design of reactor must encompass an analysis of the whole technology, considering social and engineering aspects.

REFERENCES FOR PART II

1. Choi, S.Y., Ryu, D.D.Y., Rhee, J.S. 1982. Biotechnol. Bioeng. **24**: 1165.
2. Constantinides, A. 1980. Biotechnol. Bioeng. **22**: 119.
3. Deslie, W.D., Cheyran M. 1981. Biotechnol. Bioeng. **23**: 2257.
4. Denbigh, K.G., Turner, J.C.R. 1984. *Chemical Reactor Theory. An Introduction.* 3rd Ed. Cambridge University Press, N.Y.
5. Gray, P.P., Bhuwapathanapun, S. 1980. Biotechnol. Bioeng. **22**: 1785.
6. Heertjes, P.M. 1982. Biotechnol. Bioeng **24**: 443.
7. Hirt, W. 1981. Biotechnol. Bioeng. **23**: 235.
8. Horak, J., Pasek, J. 1978. *Design of Industrial Chemical Reactors from Laboratory Data.* Heyden and Son, Ltd. London, U.K.
9. Labuza, T.P., LeRoux, Z.J.P., Fan, T.S. 1970. Biotechnol. Bioeng. **12**: 135.

BIBLIOGRAPHY FOR PART II

Fielder,R.M., Rousseau, R.W. 1978. *Elementary Principles of Chemical Processes.* J. Wiley and Sons, N.Y.

Henley, E.J., Bieber, H. 1959. *Chemical Engineering Calculations, Mass and Energy Balances.* McGraw-Hill Book Company, N.Y.

Henley, E.J., Rosen, E.M. 1969. *Material and Energy Balance Computations.* J. Wiley and Sons, N.Y.

Himmelblau, D.M. 1967. *Basic Principles and Calculations in Chemical Engineering.* Prentice-Hall, Inc., Englewood Cliffs, N.J.

Hill, G.H. Jr. 1977. *An Introduction to Chemical Engineering Kinetics and Reactor Design.* J. Wiley and Sons, N.Y.

Levenspiel, O. 1965. *Chemical Reaction Engineering* J. Wiley and Sons, N.Y.

Schmidt, A.X., List, H.L. 1962. *Material and Energy Balances.* Prentice-Hall, Inc., Englewood Cliffs, N.J.

Whitwell, J.C., Toner, R.K. 1969. *Conservation of Mass and Energy.* Blaisdell Publishing Co., Wiltham, MA.

PART III

CASE STUDY

7. MODELING OF THE ACETONE-BUTANOL-ETHANOL (A-B-E) FERMENTATION PROCESS ALTERNATIVES

Introduction

In order to demonstrate the principles discussed in this text, it would be instructional to do so in a comprehensive "case study" covering a broad selection of different basic operating modes of a suitably chosen fermentation process. The fermentation process selected here for this purpose will be the well established fermentation conversion of carbohydrate raw materials (sugar) into a dilute mixture of organic solvents, namely acetone, butanol and ethanol (A-B-E) as end-products accumulating in the fermentation broth. This anaerobic fermentation process represents a complex growth-nonassociated product formation proceeding via two growth-associated intermediates, butyric and acetic acids. In addition, a mixture of hydrogen and carbon dioxide is given off during the bioconversion catalyzed by a strict anaerobic bacterium *Clostridium acetobutylicum* which possesses a fascinating physiological behaviour being a spore former.

It is extremely difficult, if not impossible, to efficiently study, elucidate and optimize a fermentation system with a number of process variables by using a conventional empirical and experimental approach. Recent advances in the methodology of bio-process modeling, together with the accumulated knowledge of biochemistry and physiology associated with the A-B-E fermentation, make it feasible to consider an advantage which can be drawn from formulation of a mathematical model of the fermentation system. The model can assist in further explorations of the fermentation by:

- guiding the experimental work by computer simulations of process alternatives,
- providing a basis for dynamic control and optimization of the fermentation system.

A number of kinetic models of microbial systems are described in the literature[43]. However, they often represent only a very crude approximation of the actual process behaviour and variability of growth and production physiology. This is particularly the case for unsteady state culture periods, the lag and the declining growth phases. The usefulness and application potential of these models in computer simulation studies of fermentation process dynamics is invariably severely limited.

The A-B-E fermentation process selected for the "Case Study" here is sufficiently complex and can serve extremely well as an example for study and modeling of simple fermentation systems where the modeling patterns and methodology established in this "case study" can be easily followed. Different operating modes of the selected fermentation will be considered in this study which leads from a straight batch fermentation through typical continuous-flow operating modes to several basic operating modes using immobilized-cell technology. For the development of mathematical models of the above culture alternatives it is essential to consider the biochemical stoichiometry of the process, kinetics of product formation and substrate utilization, hydraulic conditions in the bioreactor and eventually even the diffusional characteristics of the chemical species through the cell immobilizing gel materials for bioreactors using entrapped immobilized cells. The resulting sets of model equations representing the fermentation system can become fairly complex and some basic guidance is given for handling the mathematical solutions. A properly identified general mathematical model of each operating fermentation process mode is demonstrated here in a simple computer simulation exercise to illustrate the expected bioreactor behaviour under selected culture conditions. The bioreactor behaviour models presented here are based on mass balances for the key chemical species of the fermentation process. This approach has a great advantage in that the individual mathematical terms used in the model equations have a real physical meaning and interpretation in the process.

For the development of the models describing the immobilized-cell process behaviour it was necessary to express the basic kinetics of the given fermentation. This had to be done for the batch culture mode where an adequate data base exists. The resulting mathematical model of the culture kinetics is also presented in this work together with an elaboration on the basic culture aspects leading to its synthesis.

The basic batch kinetics of the process had to be extended when a continuous-flow culture arrangement was considered. Once again, before the immobilized-cell fermentation process could be considered, a mathematical formulation of process mass balances had to be prepared for the culture alternative with continuous flow of the culture medium. This has also been adapted for the process case whereby the cells are retained in the bioreactor while the liquid phase is passed through it – a bioreactor configuration with suspended free cells retained within the fermentor (filtered continuous culture).

As for any model of a real system, further and much improved model alternatives can be developed. The models presented in this work can be used for computer simulations of the bioreactor behaviour. There are many alternatives of applying the models which depend on the type of computer hardware employed and calculation subroutines available for this purpose. It is recommended that anyone attempting to make use of the mathematical models presented here develops his own computer programs suited to the computer hardware and degree of sophistication of further work to be carried out.

Description of the A-B-E Fermentation Process

The discovery of the production of acetone and butanol by microorganisms was made in 1862 by Pasteur. It was in 1912 that Weizmann reported the isolation of a bacterial culture which had the capability of fermenting grains such as corn, in the absence of any additional nutrient, into acetone and butanol. The Weizmann process, as it is technically known, was quickly commercialized in 1914 as a result of the First World War demand for acetone. While a number of bacterial strains, particularly from the genus *Clostridium*, are capable of producing different types of solvents, the key strain for industrial acetone-butanol production has been *Clostridium acetobutylicum*. This bacterium and its behaviour were the focus of numerous microbiological and physiological fundamental studies during the period from 1900 to the 1950s with a renewed recent interest in the process[2,31].

Viable cells of this bacterial strain are strictly anaerobic, motile, straight rods with rounded ends occurring singly or in pairs, but not in chains. Oval, eccentric, or subterminal spores, when formed, swell the cell. The microorganism can be isolated from soil, maize, molasses, potatoes, and similar anaerobic habitats. Oxygen is toxic to the vegetative cells which require a redox potential of -250 mV or less. The cells, measuring 0.6–0.72 μm are Gram–positive when young, becoming Gram–negative after $\sim$ 60 hrs. After $\sim$ 36 hrs, the cells also lose their motility.

Clostridium acetobutylicum is capable of growth and solvent production from mashes of various grains (corn, wheat, barley, rice) and tubers (potatoes, cassava, beets) without any special nutrient additions. Additions of ammonium salts and soluble phosphorus-containing nutrients are required. On a synthetic medium, the culture demonstrated requirements for at least two growth factors, namely asparagine and coenzyme(s) available in yeast or maize extract.

Hexose and pentose-type sugars are readily fermented into acetic and butyric acid intermediates which are further converted into the three endproducts, acetone, butanol, and ethanol, with a simultaneous production of hydrogen and CO_2 gases.

A normal solvent ratio is usually 60:30:10 (butanol:acetone:ethanol) with a total solvent accumulation rarely exceeding 1.8 % in the batch culture due to the toxic effect of (mainly) the butanol fraction.

The common intermediate for the solvent biosynthesis is Acetyl-CoA derived through a complex multienzyme-catalyzed step from pyruvate which, in turn, is the end-product of the glycolysis sequence (Figure 7.1). It is evident from this figure that the control of acetate production is the key point of the whole enzymatic sequence of solvent synthesis. This is, in fact, so since all the end-product solvents are comprised of acetate units. In addition to acetate formed from glucose, some *Clostridia* are capable of conversion of CO_2 into more acetate. Respective additions of acetic, pyruvic and acetoacetic acids into the broth resulted in increased quantities of acetone formed with no effect on butanol, while the addition of butyric acid was reflected in increased amounts of butanol. Butyric acid is converted at a faster rate and hence a larger quantity of butanol is formed in the total solvents. Obviously, butanol is formed from butyric acid, while acetate is the precursor of acetone. In general, yields of solvents based on the carbohydrate substrate used range from 30 to 34%.

Hydrogen and carbon dioxide gases are produced throughout the fermentation process, the gaseous phase being richer in H_2 in the early stage of fermentation, while more CO_2 is produced later on. The final ratio of CO_2:H_2 is usually 60:40 by volume, the total gas productivity peaking during the last third of a batch process.

While the optimal pH for the growth of *C. acetobutylicum* is 6.5, solvent production is favoured at pH 4.5. Temperatures of 35-37°C are preferred. The duration of a conventional batch culture varies from 36 to 60 hours. The initial phase is characterized by cell growth and conversion of sugars into acids, accompanied by a decrease in pH from 6.0 to 4.5. The titratable acidity typically increases for up to one-half of the batch time, while the solvent synthesis commences approximately one-fourth of the way into the batch. The maximum solvent production rate corresponds to the maximum of acidity accumulation.

From the typical growth and product formation curves, it can be seen that the whole batch process consists of two phases corresponding to the two-stage product formation mechanism involved. The young cells in the acid-formation phase do not possess the enzymatic apparatus for reducing the acids. The early termination of the second phase is due to butanol toxicity and substrate depletion.

It was suggested that, due to the above reasons, it may be beneficial to carry out the process in several stages on a continuous-flow basis. A packed bed of immobilized whole cells was also examined on a laboratory scale with encouraging results. All of these culture modes will be modeled in PART III of this book showing not only the development of individual respective mathematical models but also some process simulation results pointing out ways of process optimization.

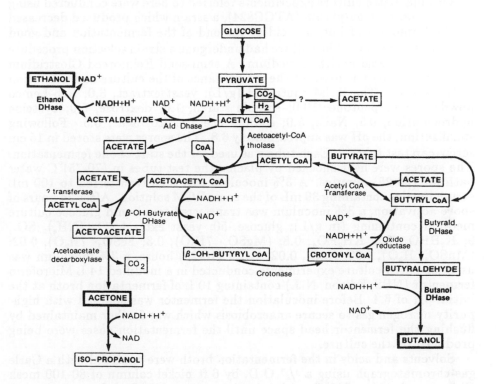

Figure 7.1: Biochemical pathway for conversion of sugar into organic solvents by *Clostridium acetobutylicum*.

7.1 BATCH CULTURE

7.1.1 Experimental

Original batch culture experiments referred to here were conducted using *Clostridium acetobutylicum* (ATCC824), a strain which produced decreased residual amounts of butyric acid at the end of the fermentation and good solvent production[2]. The culture had undergone a strain selection procedure using butyric acid enriched medium. A semi-solid Reinforced Clostridium Medium (RCM) was used for the maintenance of the culture and activation of the spores. The RCM contained (g/L): yeast extract, 3.0; Lab-Lemco powder, 10.0; peptone, 10.0; soluble starch, 1.0; dextrose, 5.0; cysteine hydrochloride, 0.5; NaCl, 5.0; sodium acetate, 3.0; agar, 0.5. Following sterilization, the pH was approximately 6.8. The spores were stored in 15 cm screw-cap test tubes[31]. Each tube was used for the start of one fermentation. The spores were heat shocked by placing the test tubes in (70-75)°C water bath for (20-25) minutes. A 3% inoculum was then transferred to 100 mL serum bottles containing 30 mL of the same RCM solution. After 24 hours of spore activation, a 3% inoculum was transferred to a liquid glucose culture medium containing (in g/L): glucose, 50; yeast extract, 11; $(NH_4)_2 SO_4$, 9; K_2HPO_4, 0.8; KH_2PO_4, 0.8; $(MgSO_4 \cdot 7H_2O)$, 0.3; $FeSO_4 \cdot 7H_2O)$, 0.02; $(MnSO_4 \cdot H_2O)$, 0.02; NaCl, 0.02. After (16-20) hours, a 3% inoculum was used for batch culture experiments conducted in a modified 14 L Microferm fermentor (NBS, Edison, N.J.) containing 10 L of fermentation broth at the initial pH of 6.4. Before inoculation the fermentor was sparged with high-purity nitrogen gas to secure anaerobiosis which was further maintained by flushing the fermentor head space until the fermentation gases were being produced by the culture.

Solvents and acids in the fermentation broth were analyzed with a Carle gas-chromatograph using a $\frac{1}{8}$" O.D. by 6 ft nickel column of 80-100 mesh Chromosorb 101 at 210°C, with injection port at 210°C. A flame ionization detector was used and the carrier gas was helium at the flow rate of 60 mL/min. Chromatographic samples were acidified by adding 0.25 mL of a 2% (w/v) sulfuric acid solution to 1 ml of cell free samples to ensure that butyrate and acetate were in the acids form. Fermentation gases were also analyzed by a gas chromatograph (Hewlett-Packard 700) using a $\frac{1}{8}$" O.D. by 12 ft SS column with 80-100 mesh Porapak Q at 70°C, with injection port at 130°C. A TC detector was used at 210°C. Nitrogen carrier gas flowed at 60 mL/min. Quantitative evaluation of chromatographic data was carried out by a Sigma 10B (Perkin-Elmer) chromatography data station. The total fermentation gas volume was determined by a wet test meter.

In addition to the dry weight biomass concentration determination, optical density of the broth was measured by a Spectronic 70 spectrophotometer (Bausch and Lomb) at 610 nm. The glucose concentration in the broth was determined enzymatically by a Fisher Diagnostic Glucose Hexokinase (HK) method.

7.1.2 Formulation of the Model

A process oriented mathematical model of the acetone-butanol fermentation has to reflect the biochemical kinetics of the process and the culture physiological aspects. Biosynthesis of solvents by *Clostridium acetobutylicum* has been elucidated in its main steps and can be described by a metabolic pathway scheme shown in Figure 7.1. The mechanism of glucose utilization by the culture, following the initial glycolytic steps, is proceeding towards the main end-products which are formed on each side of the central cycle. Butanol as the main solvent end- product is formed via a butyrate intermediate in the right-hand branch of the pathway, while acetone is produced via acetate in the left-hand branch which is also concerned with the biosynthesis of ethanol. Gaseous by-products are generated in both branches as well as in the main line of the pathway before its branching. Knowledge of metabolic sequences is essential for formulation of the mass balance equations whereby the "appearance" and "disappearance" of individual intermediates and products in the bioreactor system are quantitatively accounted for.

The physiology of the culture is less well understood and only apparent physiological features of the fermentation process have been described. Production strains of *Clostridium acetobutylicum* are characterized by a relatively long lag phase, variable morphology during cultivation, pronounced sensitivity to pH, and by a strong inhibition of growth and metabolic activities by butanol[18] accompanied by cell lysis, or even sporulation. Based on the concept of the physiological state of the culture[16], and on the von Bertalanfy[48] systems theory aimed at biological applications, a conclusion has been made for an adequate mathematical description of the A-B-E process. A structured growth model[12] has to be capable of at least partially describing the variability of the growth dynamics which depends on the history of the microbial culture. A suitably chosen marker of the physiological state can serve to introduce this unconventional aspect into the process model. A morphological image of the microbial population[30] or a concentration of some intracellular component[7] which significantly changes during the growth can be used as such a marker. The concept developed by Powell[23] in his theoretical work dealing with modeling of transient states in the microbial culture has been introduced in Section 2.3 of this text. It represents an interesting possibility of modeling the relationship between the specific growth rate μ, the culture history and variable environment using the "metabolic activity functional" Q. This relationship is defined as:

$$\mu = Y_{x/s} \, Q \, g(S) \tag{7.1}$$

where $Y_{x/s}$ is the theoretical, thermodynamically maximum, macroscopic yield coefficient[28], and $g(S)$ is a simple function depending on the environment.

It can be used for simulation of transient states of the culture which result from a perturbation imposing a stress upon the microbial population[42]. In case of the fermentation considered here, it may be for example the loss of anaerobic culture conditions, a substrate shock, or another kind of perturbation resulting in destabilization of the culture physiological equilibrium. For the experimental data from the A-B-E process which are used for model identification in this work, the value of $g(S)$ function can be considered as unity.

The metabolic activity functional $Q(t)$ is not a simple function of time but its value depends on the culture history and the variable substrate consumption rate during different stages of development of the microbial population. This functional has been defined by Powell[23] as:

$$Q(t) = \int_0^\infty f(\xi) \; q \; [S(t - \xi)] \; d\xi \tag{7.2}$$

where $f(\xi)$ is a distribution function characterizing the structure of the population according to age, q is the specific substrate consumption rate for a certain age category.

The intracellular RNA concentration can be readily used as a marker of the culture physiological state because RNA concentration exhibits a linear relationship with the cell growth rate, and because the ratio of individual RNA sub-components (mRNA, tRNA, rRNA) is usually constant over a broad range of culture conditions.

The relationship between the growth rate and the concentration of the intracellular marker of the culture physiological state (e.g. RNA) can be expressed as follows:

$$\mu = const \, (\text{RNA} - \text{RNA}_{min}) \, g(S) \tag{7.3}$$

where RNA_{min} is the RNA concentration in the cell at $\mu = 0$.

This basic relationship has to be supplemented by a dynamic mass balance for the intracellular RNA. Considering the A-B-E process, it was experimentally established that the culture growth rate is directly proportional to the substrate (sugar) concentration multiplied by a term characterizing the inhibition of culture growth by butanol. Within a certain concentration range this represents an allowable simplification of the part of the Monod kinetics model where the K_S is comparable to the initial sugar concentration in the batch. For the batch culture model the parametric sensitivity analysis reveals that the model is not profoundly influenced by this simplification due to its low sensitivity to K_S. In comparison, the same parametric analysis indicates that the sensitivity to the K_I is higher by an order of magnitude, since the concentration of butanol represents a strong negative metabolic feed-back.

$$\frac{\text{dRNA}}{\text{d}t} \, X = k_1 S \, \frac{K_I}{K_I + B} \, (\text{RNA} \; X) \tag{7.4}$$

The dimensionless concentration of RNA, designated as y, was used in this work as a marker of the culture physiological state whereby

$$y = \frac{RNA}{RNA_{min}} \tag{7.5}$$

This way the specific growth rate of the culture which has been shown to be related to the cellular RNA content[12] can be expressed as

$$\mu = ay - b \tag{7.6}$$

Harder and Roels[12] demonstrated that the numerical values of the above coefficients are constant for most bacterial cultures. Consequently, a parameter λ can be defined as

$$\lambda = const \; \frac{g(S)}{RNA_{min}} = 0.56 \tag{7.7}$$

which characterizes numerous bacterial cultures. This approximation allows the use of this concept in the present model. The RNA content of the culture was not actually determined in this work but it was deemed desirable to build this aspect into the model which becomes potentially more responsive to the changes in the culture physiology as indicated by the "marker" compound.

Evaluation of the culture dynamics associated with the marker of the culture physiological state can be performed also for *C. acetobutylicum* by expressing the following dimensionless differential balance:

$$\frac{d(yX)}{dt} = \mu(S, B) \cdot y \cdot X \tag{7.8}$$

where function $\mu (S,B)$ is a characteristic of the culture dependent on respective concentrations of the limiting substrate S, and the inhibitory product, B. Equation (7.8) can be mathematically rearranged to assume the following form:

$$\frac{dy}{dt} = \mu(S, B)y - 0.56(y - 1)y \tag{7.9}$$

The initial condition for $y(0)$ is $y(0)=1$ which characterizes the inoculum in its stationary phase. An inoculum in its exponential phase of growth would be characterized by $y(0) \geq 1$, while an inoculum in its declining phase, when only a part of biomass is capable of growth, would be characterized by $y(0) \leq 1$. This initial condition characterizes the physiological state of the culture with regard to the previous culture history during the inoculum propagation. This approach also describes the viability of the culture at the beginning and enables a better representation of the process in question by a mathematical model. For the extent of computer simulation of the acetone-butanol fermentation considered in this work it was not crucial to actually determine the RNA levels in the biomass and the approximation suggested by Harder and Roels[12] was conveniently used instead.

The function $\mu(S,B)$ can be identified from the experimental data. For the A-B-E process, a linear relationship with respect to substrate is combined with the simultaneous product inhibition of the Yerusalimski-Monod type as seen in Equation (7.4). In expressing the differential mass balance for the biomass, the cell decay and lysis has to be considered which is directly proportional to the concentration of butanol (B) in the broth, the final equations assuming the following forms:

$$\frac{dX}{dt} = \lambda(y-1)\,X - k_2 XB \tag{7.10a}$$

$$\frac{dy}{dt} = [k_1 S\,\frac{K_I}{K_I + B} - \lambda(y-1)]y \tag{7.10b}$$

The changes in the physiological activities of *C. acetobutylicum* were further evaluated by studying the kinetics of the A-B-E process. The observed specific rate of change of a process variable is caused by the internal physiological functions of the cell which are reflected in the process kinetic, it is usually designated as q. It is customary to use r to designate a mathematical approximation of the rate of change in the accumulation of the considered mass balance species. Relationships were studied between specific rates and concentrations of appropriately selected fermentation parameters characteristic of the process kinetics. Evaluation of the rate-concentration relationship was based on graphical plots of these quantities. The original results[52] of four pH-controlled batch fermentations were utilized for this purpose. The fermentation variables considered were the end-product solvents (acetone, butanol, ethanol), gases (carbon dioxide and hydrogen), intermediates (butyric and acetic acids) and the self-propagating biocatalytic bacterial biomass. The corresponding observed specific rates of change of these chemical fermentation process variables are listed below:

X, biomass concentration   specific growth rate $\mu = \dfrac{1}{X}\,\dfrac{dX}{dt}$

S, substrate concentration   specific substrate utilization rate $q_S = \dfrac{-1}{X}\,\dfrac{dS}{dt}$

B, butanol concentration   specific rate of butanol production $q_B = \dfrac{1}{X}\,\dfrac{dB}{dt}$

BA, butyrate concentration   specific rate of $q_{BA} = \dfrac{1}{X} \dfrac{\mathrm{d}BA}{\mathrm{d}t}$
                                 butyrate production

A, acetone concentration   specific rate of $q_{A} = \dfrac{1}{X} \dfrac{\mathrm{d}A}{\mathrm{d}t}$
                                 acetone production

AA, acetate concentration   specific rate of $q_{AA} = \dfrac{1}{X} \dfrac{\mathrm{d}AA}{\mathrm{d}t}$
                                 acetate production

E, ethanol concentration   specific rate of $q_{E} = \dfrac{1}{X} \dfrac{\mathrm{d}E}{\mathrm{d}t}$
                                 ethanol production

CO_2, carbon dioxide   specific rate of
    concentration   carbon dioxide $q_{CO_2} = \dfrac{1}{X} \dfrac{\mathrm{d}CO_2}{\mathrm{d}t}$
                                 production

H_2, hydrogen concentration   specific rate of $q_{H_2} = \dfrac{1}{X} \dfrac{\mathrm{d}H_2}{\mathrm{d}t}$
                                 hydrogen production

in general:

P, product concentration   specific rate of $q_{P} = \dfrac{1}{X} \dfrac{\mathrm{d}P}{\mathrm{d}t}$
                                 product production

7.1.3 Smoothing of Experimental Data

Experimental data available from selected A-B-E batch fermentation experiments need to be smoothed to facilitate further use of the information. In general, the following relationship is dealt with which was discussed in Section 3.1

$$y_i(t) = f_i(t) + \epsilon_i \qquad (7.11)$$

where y_i is the measured value,
        ϵ_i is the measuring error, and
        f_i is the approximating function.

The experimental measurements were done in a time series t_1; t_2; $\cdots$ t_N with a corresponding series of experimental values y_{i1}; y_{i2}; $\cdots$ y_{iN}. The estimate of the measuring error ϵ_i can be independently done based on several known formulae. In case of the fermentation process considered, the estimate of this error value is in the range 0.1 - 0.5 of the absolute experimental value for each independent variable. For smoothing of the experimental data it is possible to use some of the approximation methods available in the literature. However, for the best curve fitting and for the most accurate estimate of the numerical value for the dependent variable differential $y_i(t)$ with t, it is advisable to use a piecewise polynomial approximation method based on the theory of so called "spline functions". Details of this approach can be found, for example, in the work of deBoor[3]. While the methodology used here for evaluation of specific velocities of individual fermentation reactions is also based on the above work, the approximation of experimental data uses the SMOOTH[3] complex of programs to minimize the following function over all the values of the approximating polynomial and its m-th derivative:

$$p\sum_{i=1}^{N}[\frac{y_i(t) - f_i(t)}{\epsilon_i}]^2 + (1 - p)\int_{t_1}^{t_N}[f^m(\tau)]^2\,d\tau \qquad (7.12)$$

This approach to minimizing the above relationship represents a compromise between the requirement for the approximation curve to be the closest to the experimental data on one hand and for the relationship to be smoothest on the other. This choice is determined by the magnitude of ϵ_i estimate which, in turn, automatically determines the value of weight coefficient p. Again further details and description of the algorithm is available in the above mentioned work of deBoor[3] (Chapter XIV). The Computer program published in his book was transcribed from computer language FORTRAN 77 into FORTRAN IV and has been de-bugged for the use in the interactive computer system of McGill University (MUSIC) in Montreal, Canada.

For the experimental data analysis, program SMOOTH, CHOLID and SETUPQ were used, supplemented by the main program enabling data input according to the block diagram in Figure 7.2. Simultaneous transformation and smoothing of the results was facilitated through the use of subroutine TRANSF. The smoothed values of dependent variables and their first derivatives are used in this program according to the users demand for calculations concerning specific individual reactions in the fermentation process. Particularly important for the subsequent formulation of the process model are the culture specific growth rate μ, the substrate utilization and the product synthesis rates q_S and q_P respectively.

Table 7.1 (in the Appendix) presents a listing of the computer program for smoothing of experimental data and an example of a TRANSF subprogram for analyzing batch culture data. Smoothed derivatives of dependent variables in vector $Y(I,J)$ and the corresponding values of specific rates after the calculation are stored in vector F. Variables YD and Y are organized in such a way that index I indicates the sequence along the time axis $(t_1; t_2; \cdots t_N)$ and index J indicates the sequence of input data. TRANSF can be used in future also for statistical data processing using BMDP, MINUIT, SAS and other software means of mathematical statistics.

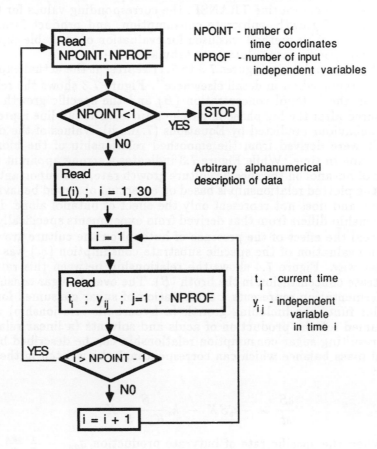

Figure 7.2: Block diagram of data input for BIOKIN-A computer program capable of modeling and simulation of the bioreactor behavior.

For better illustration and eventual further application of the program by other users, Table 7.2 (Appendix) shows an example of the program application and a fermentation experiment input data format in the following sequence: time, concentrations of biomass, glucose, butanol, acetone, ethanol, butyric acid, and acetic acid, respectively. At the same time the program output is also shown.

Following the translating, the input data print-out check is carried out. By using the subroutine TRANSF, the corresponding values for time, specific rates of growth, substrate consumption, and product formation are printed out. This method was used for evaluation of available experimental data and specific rate values were tabulated.

The data points in Figures 7.3 to 7.11 represent the actual experimental data as reported on in detail elsewhere[52]. Figure 7.3 shows the relationship between the butanol concentration (B) and the specific growth rate μ as measured after the lag phase in batch cultures. The full line represents culture behaviour predicted by Equations (7.10a,b). Values of the differential dX/dt were derived from the smoothed relationship of the biomass concentration in time. While Figure 7.3 indicates a strong apparent inhibitory effect of butanol on the specific culture growth rate, it is important to realize that the plotted relationship is based on an actual observed behaviour of the process and does not represent only the effect of butanol alone. In this the relationship differs from that derived from experiments specifically designed to reveal the effect of the presence of butanol on the culture growth[18].

An evaluation of the specific substrate consumption (q_S) was done in a similar way. Figure 7.4 shows the relationship between this rate and the substrate concentration in the broth (S). The overall sugar consumption in the fermentation represents a combination of sugar consumed for different cellular functions including growth (a hyperbolic relationship) and sugar consumed for the production of acids and solvents (a linear relationship). The resulting sugar consumption relationship can be described by a differential mass balance which can correspondingly be written in the following form:

$$\frac{dS}{dt} = -k_3 SX - k_4 \frac{S}{K_S + S} X \tag{7.13}$$

When the specific rate of butyrate production $q_{BA} = \frac{1}{X} \frac{dBA}{dt}$ is plotted against the concentration of butanol a very similar relationship is obtained (see Figure 7.5) to the one expressed between μ and b. The negative values of q_{BA} represent the metabolic activity where the production of butyrate is inhibited by butanol and the conversion of butyrate into butanol is taking place. A butyrate mass balance for the reaction system can be expressed in the following form:

$$\frac{dBA}{dt} = k_5 S \frac{K_I}{K_I + B} X - k_6 \frac{BA}{K_{BA} + BA} X \tag{7.14}$$

Figure 7.3:

Relationship of the specific growth rate μ [h⁻¹] and the broth butanol concentration in the bioreactor (*B*) for a batch culture of *Clostridium acetobutylicum.*

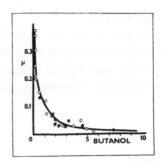

Figure 7.4:

Relationship of the specific substrate utilization rate q_Z [g glucose/h.g biomass] and the broth substrate concentration (*S*) for a batch. culture of *Clostridium acetobutylicum.*

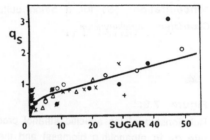

Figure 7.5:

Relationship of the specific butyric acid production rate q_{BA} [g butyric acid/h.g biomass] and the broth butanol concentration (*B*) for a batch culture of *Clostridium acetobutylicum.*

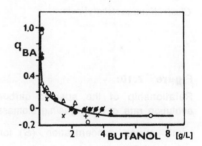

Figure 7.6:

Relationship of the combined specific production rates of the culture inhibitory substances (q_B + 0.841 q_{BA}) and and the broth sugar concentration (*S*) for a batch culture of *Clostridium acetobutylicum.*

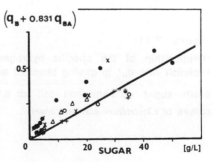

Figure 7.7:

Relationship of the specific acetic acid production rate q_{AA} [g acetic acid/h.g biomass] and the broth butanol concentration (B) for a batch culture of *Clostridium acetobutylicum*.

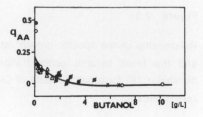

Figure 7.8:

Relationship of the combined specific production rates ($q_A + 0.484\ q_{AA}$) and the broth sugar concentration (S) for a batch culture of *Clostridium acetobutylicum*.

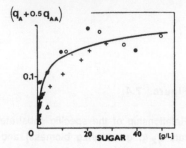

Figure 7.9:

Relationship of the specific ethanol production rate q_E [g ethanol/h.g biomass] and the broth sugar concentration (S) for a batch culture of *Clostridium acetobutylicum*.

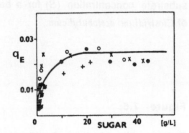

Figure 7.10:

Relationship of the specific carbon dioxide evolution rate q_{CO_2} [g gas/h.g biomass] and the broth sugar concentration (S) for a batch culture of *Clostridium acetobutylicum*.

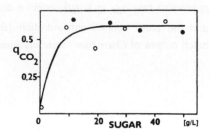

Figure 7.11:

Relationship of the specific hydrogen gas evolution rate q_{H_2} [g gas/h.g biomass] and the broth sugar concentration (S) for a batch culture of *Clostridium acetobutylicum*.

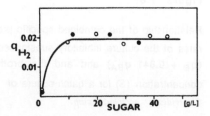

In this mass balance the first right-hand side term represents biosynthesis of butyrate from sugar substrate inhibited by butanol. The second one reflects the consumption of butyrate for its bioconversion into butanol. K_I and K_{BA} are inhibition and saturation constants, respectively, for the two reactions. A subsystem consisting of butanol and butyric acid was considered for the analysis of butanol production kinetics. When the stoichiometrically converted rate of butyrate production, i.e., 0.841 q_{BA}, is added to the butanol production rate $q_B = \frac{1}{X} \frac{dB}{dt}$, a linear relationship could be assumed when plotted against the substrate concentration S. In order to maintain a stoichiometric consistency for this process this assumption could be made based on the sum of yields for all the products which has to be maintained constant. This way of expressing this relationship reflects the existence of invariant between the consumption of sugar in the right hand branch of the metabolic pathway leading to the synthesis of butanol via butyric acid. It is essential to stress that an implicit assumption has also been made concerning the biomass synthesis which has been assumed to be proportional to the consumption of yeast extract from the medium[53]. The result of these considerations is a plot seen in Figure 7.6. Overall, a linear relationship is assumed. The differential mass balance for butanol in the system, which enables simulation of the initial delay in the butanol production and accumulation caused by the intermediate accumulation of butyrate in the broth, can be expressed in the following form:

$$\frac{dB}{dt} = k_7 SX - 0.841 \frac{dBA}{dt} \tag{7.15}$$

Coefficient 0.841 resulted from the stoichiometric conversion considerations as a ratio of molecular weights of butanol and butyric acid. It was already mentioned earlier that butanol has a pronounced inhibitory effect on the reactions concerned with growth and synthesis of cellular material which is in turn closely related to the production of butyrate.

A similar controlling effect of butanol is also seen in Figure 7.7 showing the relationship of the specific rate of acetate synthesis (q_{AA}) and butanol concentration B. The Figure indicates that this relationship is very similar to that for butyrate (Figure 7.5). The mass balance for acetic acid can be written as:

$$\frac{dAA}{dt} = k_8 \frac{S}{K_S + S} \frac{K_I}{K_I + B} X - k_9 \frac{AA}{K_A A + AA} \frac{S}{K_S + S} X$$

Similarly, as for kinetics of butyrate production expressed in Equation (7.12), the first term on the right side of mass balance Equation (7.14) represents the rate of acetate biosynthesis while the second one is the rate of acetate conversion into acetone.

When the stoichiometrically converted rate of acetate biosynthesis, i.e., 0.484 q_{AA}, is added to the specific rate of acetone production q_A and the sum of the two is plotted against the substrate concentration S, a curve results which corresponds to the rate of sugar consumption described by the second term in the sugar balance expressed by Equation (7.11). The relationship for the combined rates and S is shown in Figure 7.8. The dynamics of acetone production is then described analogically to that for the butanol mass balance in Equation (7.13) as follows:

$$\frac{dA}{dt} = k_{10} \frac{S}{K_s + S} X - 0.484 \frac{dAA}{dt} \tag{7.17}$$

where

$$k_{10} \frac{S}{K_s + S} = q_A + 0.484 \, q_{AA}$$

Upon examination of e.g. Fig. 7.6 and Fig. 7.8 the scatter of experimental data points may seem somewhat broad. It is essential to bear in mind, however, that only one set of experiments was used for this purpose in which experimental conditions varied. Also, the equations consider and are capable of describing a well established general culture behaviour of C. *acetobutylicum* in batch culture as reported by many studying the process.

For the complete description of the solvent biosynthesis by C. *acetobutylicum*, the ethanol mass balance remains to be expressed. When the specific rate of ethanol production (q_E) is plotted against the sugar concentration S a relationship is observed from a batch culture such as depicted in Figure 7.9.

For simulation of the ethanol production rate in the A-B-E process the Monod-type function can be used with an adequate degree of accuracy:

$$\frac{dE}{dt} = k_{11} \frac{S}{K_s + S} X \tag{7.18}$$

A similar approach can also be taken for modeling the relationship of the rate of evolution of fermentation gases (CO_2 and H_2) and the substrate concentration. Figures 7.10 and 7.11 present respectively the specific evolution rate of carbon dioxide and hydrogen [g(gas)/g(d.w.)·h] and as apparent from these diagrams, the following expressions are adequate for describing the depicted relationships:

$$\frac{dCO_2}{dt} = k_{12} \frac{S}{K_s + S} X \tag{7.19a}$$

$$\frac{dH_2}{dt} = k_{13} \frac{S}{K_s + S} X + k_{14} \, SX \tag{7.19b}$$

The system of differential mass balances expressed by Equations (7.10a, b) to (7.17a,b) represents a general mathematical model capable of describing dynamics of the A-B-E process based on a batch fermentation.

7.1.4 Model Parameter Identification and Parametric Sensitivity

In order to be able to use the general mathematical model developed in the preceding section e.g. for computer simulation studies of the bioreactor performance and to verify its capability to describe the fermentation dynamics, numerical values of model parameters k_1 to k_{14} used in Equation (7.10) to (7.17) have to be determined. This model identification procedure is based on minimizing the deviations between the model predictions and actual experimental data. A non-linear regression technique[1] can be used for this purpose in combination with a set of computer programs called BIOKIN[50] which enables a simultaneous evaluation of fermentation kinetic parameters.

An appropriate criterion for the fitting of the model could be, among others, the Euclidean norm of distance given by the sum of squares of individual deviations between the model and the data:

$$SSR = \sum_{j=1}^{M} \sum_{i=1}^{N} (Y_{ij_{model}} - Y_{ij_{experiment}})^2 \qquad (7.20)$$

Summation of the residues is done over a number of experimental points $(i = 1,2...M)$ and over a number of measured dependent variables in one experimental point $(j = 1,2...N)$. When the measured values of dependent variables differ by an order of magnitude in their numerical value, it is appropriate to use the sum of squares of weighed residues for the data fitting. In this approach the magnitudes of the measured quantities are normalized by an appropriate weighing coefficient (usually this is a square of the reciprocal maximum numerical value of an independent variable) according to the following relationship:

$$SSW = \sum_{j=1}^{M} W_j \sum_{i=1}^{N} (Y_{ij_{model}} - Y_{ij_{experiment}})^2 \qquad (7.21)$$

The coefficient (parameter) estimation is then done as a standard task in locating a numerical extreme according to the algorithm described by a block diagram in Figure 7.12.

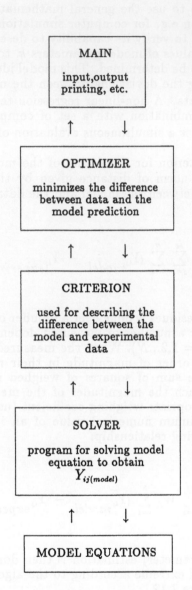

Figure 7.12: Block diagram for computer data fitting by non-linear regression in ordinary differential equation (ODE) problems.

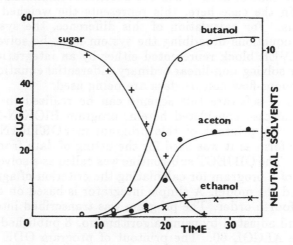

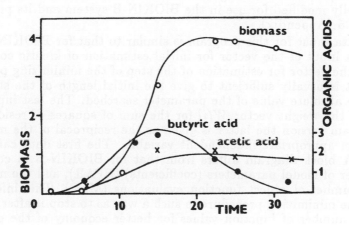

Figure 7.13 (a and b): Experimental data points and model predictions (full lines) for the main process parameters in the batch culture acetone-butanol fermentation process of *Clostridium acetobutylicum*.

In the main program (MAIN) the data input consists of the experimental data and the control data. Then the automatic minimizing of the criterion is called. In the case here, this represents the weighed sum of squares of deviations. For calculation of this difference, the system of ordinary differential equations describing the system must be solved. This is done by the SOLVER block represented either by an integration program or a program for solving non-linear ordinary differential equations when steady-state continuous-flow culture data are being used.

In terms of software this scheme can be realized in several different ways. In the case considered herein, program BIOKIN-B[50] can serve as an example. A printout of this program in FORTRAN is in Table 7.3 in the same form as it was used for the fitting of data from the cultivation experiment. The OBJECT sub-routine was called as a solver which assumed here a role of a program for calculating the criterion of agreement between the data and the model. A simple integrator is based on the Runge-Kutta method of fourth order. The program was transcribed into FORTRAN IV language and adjusted by using algorithm no. 8 published in Com. ACM[21] in language ALGOL 60. The printout of program ODE is seen in Table 7.4 (Appendix). As an optimizing program for direct search of a non-linear function minimum, the one published in the book by Kuester and Mize[14] was adopted, based on the original method by Rosenbrock[29]. This program was partially modified for use in the BIOKIN-B system and its printout is in Table 7.5 (Appendix).

The data input for this program is similar to that for BIOKIN-A[49], including an input of the vector for initial estimation of kinetic coefficients $A(i)$ and the vector for estimation of the step of the minimizing procedure EPS(i). It is usually sufficient to give the initial length of the step as $1/3$ - $1/2$ of the absolute value of the parameter searched. The last input to be entered is the weight vector $W(i)$ for the sum of squares of residuals. In this program version the latter is given as the reciprocal of the maximum value of an appropriate independent variable. The first data card in the BIOKIN-A block diagram differs from that for BIOKIN-B in containing the number of model parameters (coefficients) (NPAR), and the maximum allowed number of object function evaluations (MFC). This information governs the minimizing procedure in such a way as to stop it after reaching the given number of function values for better economy of the computer time. The result printout of each iteration contains vector of parameters A_i and an appropriate value of the criterion for agreement between the experimental data and the model. It is illustrated in an example of a BIOKIN-B program run in Table 7.6 (Appendix).

During a multiple use of this very simplified version of BIOKIN program complex, the values of model coefficients were determined which are summarized in Table 7.7. This table, together with corresponding differential balances for the substrate, biomass and products (Equations 7.10 to 7.17), represents the identified model of the acetone-butanol fermentation process.

TABLE 7.7

Values of Model Parameters (Coefficients)
and their Parametric Sensitivities
for the Mathematical Model of the Batch
Acetone-Butanol-Ethanol Biosynthesis

Parameter	Units	Absolute parametric sensitivities	Relative parametric sensitivities
$k_1 = 0.0090$	(L/g h)	202827.0	11.54
$k_2 = 0.0008$	(L/g h)	2789.9	0.014
$k_3 = 0.0255$	(L/g h)	11114.0	1.79
$k_4 = 0.6764$	(1/h)	277.49	1.187
$k_5 = 0.0136$	(L/g h)	129034.0	11.097
$k_6 = 0.1170$	(1/h)	-943.83	0.698
$k_7 = 0.0113$	(L/g h)	-99973.1	7.144
$k_8 = 0.7150$	(1/h)	30.349	0.137
$k_9 = 0.1350$	(1/h)	-63.25	0.054
$k_{10} = 0.1558$	(1/h)	-32.34	0.0318
$k_{11} = 0.0258$	(1/h)	32.17	0.00525
$k_{12} = 0.6139$	(1/h)	33.01	0.128
$k_{13} = 0.0185$	(1/h)	7.92	0.00093
$k_{14} = 0.00013$	(L/g h)	140.85	0.00011
$K_I = 0.833$	(g/L)	272.43	1.435
$K_S = 2.0$	(g/L)	0.0358	0.000453
$K_{BA} = 0.5$	(g/L)	102.95	0.3255
$K_{AA} = 0.5$	(g/L)	3.34	0.0106

TABLE 7.8

The Results of the "F-TEST"* with
the Significance (α) of 0.05 for
the Agreement Between the Experimental
and the Simulated Batch Culture Results

	S_r^2	S_e^2	S_r^2/S_e^2
Biomass	0.164	0.077	$2.18 \leq 3.0717$
Sugar	9.17	5.14	$1.78 \leq 3.0717$
Butanol	1.85	0.65	$2.84 \leq 3.0717$
Acetone	0.137	0.086	$1.59 \leq 3.0717$
Ethanol	0.035	0.012	$2.91 \leq 3.0717$
Butyric acid	0.23	0.102	$2.25 \leq 3.0717$
Acetic acid	0.114	0.045	$2.53 \leq 3.0717$
Hydrogen	0.0025	0.0018	$1.39 \leq 3.0717$
Carbon dioxide	1.24	1.39	$0.89 \leq 3.0717$

S_r = Measure of deviations around the regression function.
S_e = Measure of dispersion caused by experimental error.
* The level of significance (α) considered was $\alpha = 0.05$.

Figure 7.13a,b presents an illustrative example of a fitted relationship (full line) and experimental data. Other batch cultivations carried out as part of this work exhibited a similar pattern for the simulated curve and experimental data.

In addition to the estimation of numerical values of model coefficients, the mathematical model identification process includes determination of validity of the model for a given set of experimental data. A statistical F-test[13] was performed for this purpose, with a level of significance α = 0.05 selected for all the batch fermentation experimental data sets and key process variables. The results of the F-test presented in Table 7.8 indicate an agreement between the experimental and the simulated results for all the nine process variables.

7.1.5 Parametric Sensitivity of the Batch Culture Model

The degree of significance of individual model parameters was evaluated by the method of parametric sensitivity analysis[4]. The Absolute Parametric Sensitivity (APS), characterizing the direction in which a selected parameter is acting, is calculated as:

$$\text{APS} = \frac{\partial f}{\partial k_1} \approx \frac{\Delta f}{\Delta k_i} \tag{7.22}$$

where f is the optimized function and k_i is the parameter. Since the sensitivity cannot be analytically expressed for non-linear dynamic models, it is appropriate to use a differential approximation which can be calculated when the process simulation model is applied. The APS characterizes the direction in which the considered parameter is acting. Its positive value leads to increased difference between the model and the experimental data, and vice versa.

For a mutual comparison of the calculated parametric sensitivities and for indication of the significance sequence of individual parameters, the Relative Parametric Sensitivity (RPS) is expressed, defined as follows:

$$\text{RPS} = \left| \frac{k_i \partial f}{f \partial k_i} \right| = \left| \frac{k_i \Delta f}{f \Delta k_i} \right| \tag{7.23}$$

Similarly as for the Absolute Parametric Sensitivity, a differential approximation (7.23) can also be used for expressing the RPS.

Table 7.9 (Appendix) shows the computer program printout using all the software means mentioned previously. This program was used for calculating values of the APS and RPS for the model coefficients k_1 to k_{14}. The respective APS and RPS values are also summarized in Table 7.7. It is seen that coefficients k_1, k_5, and k_7, characterizing the kinetics of biomass growth, butanol, and butyric acid production, respectively, exhibit the highest sensitivity and can be considered as the most important parameters in the process. For further work with the A-B-E process model and for optimization of the fermentation itself, these parameters indicate the focal points.

7.1.6 Conclusion

A mathematical model of the batch A-B-E fermentation process presented in this work consists of a set of differential equations representing bioreactor mass balances for the substrate, biomass, metabolic intermediates and key end-products. The model is capable of reflecting well all batch culture phases by incorporating a culture physiological state marker. The sensitivity for the physical culture parameters such as pH, redox potential, temperature, etc., has not been incorporated in the model. Identification of the model parameters and model synthesis is based on the rate analysis of original batch experimental data which were obtained under the conventionally optimal conditions of the physical parameters and of the optimal fresh fermentation medium composition. Development of the model was based on the following basic assumptions:

- there are no culture limitations by nitrogen, phosphate, growth factors and trace elements.
- functional relationships are valid for the glucose concentration range of 0 - 50 g/L and the biomass concentration of 0.03 - 10 g/L.
- product concentrations do not exceed the limits of: butanol 11 g/L, acetone 5 g/L, ethanol 1.5 g/L, organic acids 5 g/L.
- sugar (glucose) is the only limiting substrate in the batch cultivations.

The use of numerically differentiated experimental batch culture data is appropriate only for a semi-quantitative evaluation of estimated kinetic parameters since the error in determining the derivative of a process parameter varying in time can easily exceed 50%, particularly during the initial phase of cultivation. The value of kinetic parameters determined from analyzing Figures 7.3 to 7.11 were considered only as initial estimates for iteration algorithms derived from a non-linear regression[1]. In this case, a set of computer programs called BIOKIN[50], described in an earlier work, was used for a simultaneous evaluation of kinetic parameters derived from four batch cultivation experiments carried out at different mixing rates.

The mathematical kinetic model of the batch A-B-E fermentation presented here was found to adequately represent the process, and the parametric sensitivity analysis carried out indicated the most important parameters. The model predictions in computer process simulations could be used to direct further experimental work and bioreactor performance optimization in different unconventional operating modes. The basic concept used for the batch fermentation model developed here will be used for unconventional fermentation process variants such as fed-batch, submerged continuous-flow[47] as well as immobilized-cell[51] fermentation systems which will be dealt with in the following Sections.

7.2 CONTINUOUS-FLOW CULTURE SYSTEMS

7.2.1 THE THROUGHPUT CONTINUOUS-FLOW CULTURE SYSTEM

a) *The model development*

The development of the model is based on the kinetic principles dealt with in the previous section on the batch system model development. However, slight modifications need to be introduced at this stage since under the continuous-flow culture regime, approximation of kinetic relationships used for the batch culture system, may not be adequate. In this context, a linear approximation of the relationship between the culture specific growth rate and the substrate concentration is replaced by a more appropriate and well established Yerusalimski-Monod relationship which also takes into consideration the inhibition effect of butanol in the fermentation broth. The overall functional relationship for culture growth can correspondingly be represented as:

$$\mu\,(S,B) \;=\; k_1' \,\frac{S}{S + K_s}\, \frac{K_I}{K_I + B}$$

This relationship better describes the growth kinetics in the continuous-flow culture system with values of S usually in the lower portion of the range considered for sugar concentrations. Similarly, the concentration of butanol is also not likely to reach its higher values due to the effect of continuous dilution of the broth. The relationship for $\mu(S,B)$ is determined as a characteristic of the culture and it can be expressed from the experimental data.

To describe the direction of the physiological culture activity a suitable marker of physiological state can also be used as recommended by Powell[23]. The cellular concentration of RNA was successfully used as such in the development of the batch mathematical model. Taking advantage of the findings published by Harder and Roels[12] who concluded that the relationship between the RNA concentration and the growth rate is approximately the same for most bacterial cultures, the growth rate of the culture can be expressed as:

$$\mu \;=\; 0.56\,(y\text{-}1)$$

whereby a dimensionless ratio of RNA concentrations is used as introduced in Section 2.3:

$$y \;=\; RNA\,/\,RNA_{min}$$

The dimensionless differential balance for the cellular RNA will then be

$$\frac{dy}{dt} = [k'_1 \frac{S}{S + K_s} \frac{K_I}{K_I + B} - 0.56 \, (y - 1)] \, y \qquad (7.24)$$

This is the first equation of the mathematical model for the continuous-flow throughput culture system. The remaining equations are based on other mass balances as follows:

Material balance for growth:

$$\frac{dX}{dt} = 0.56(y - 1)X - k_2 BX - DX \qquad (7.25)$$

$$\frac{dy}{dt} = [k_1 \frac{S}{S + K_s} \frac{K_I}{K_I + B} - 0.56(y - 1)] \, y \qquad (7.26)$$

Material balance for sugar substrate with feed-stream concentration S_o :

$$\frac{dS}{dt} = -k_3 SX - k_4 \frac{S}{S + K_s} X + D(S_o - S) \qquad (7.27)$$

Material balance for butyric acid:

$$\frac{d(BA)}{dt} = k'_5 \frac{S}{S + K_s} \frac{K_{IBA}}{K_{IBA} + B} X - D(BA) \qquad (7.28)$$

Material balance for acetic acid:

$$\frac{d(AA)}{dt} = k_8 \frac{S}{S + K_s} \frac{K_I}{K_I + B} X - D(AA) \qquad (7.29)$$

Material balance for butanol:

$$\frac{dB}{dt} = k'_7 \frac{S}{S + K_s} X - 0.831 \left[\frac{d(BA)}{dt} + D \cdot (BA) \right] - DB \qquad (7.30)$$

Material balance for acetone:

$$\frac{dA}{dt} = k_{10} \frac{S}{S + K_s} X - 0.5\left[\frac{d(AA)}{dt} + D \cdot (AA)\right] - DA \qquad (7.31)$$

Material balance for ethanol:

$$\frac{dE}{dt} = k_{11} \frac{S}{S + K_s} X D E \qquad (7.32)$$

The mathematical model for the continuous-flow culture system is based on the kinetics discussed and used in the development of the batch system model. However, the mass balances for the key process parameters have to respect the fact that the bioreactor with a constant working volume of broth V is subject to a continuous flow throughput characterized by the dilution rate $D = f/V$. Data for material balances concerning the gaseous phase were not determined and since they are not essential for the model identification the gas phase was not considered in this model version. An estimate for gas evolution rates can be done by using Equations (7.19a) and (7.19b) identified for the batch culture mode. The results of the model parameter identification are summarized in Table 7.10.

This model was identified on the experimental data by Leung[15] (summarized in Figures 7.14 and 7.15) who worked with a slightly different production strain due to its somewhat altered cell wall. This demonstrated itself in the fact that the apparent Monod growth constant (K_s) was lower by an order of magnitude, making it thus more comparable with the one characteristic for the left branch of the metabolic pathway. Also, the apparent K_I for the Leung's strain was several times higher than the one determined for the batch culture strain used by Yerushalmi[52]. These facts of some culture strain differences were reflected in the different format for expressing the RNA content of the culture (dy/dt) which uses the original Yerusalimski-Monod kinetics rate expression including the combined substrate limitation and product inhibition terms.

Considering that the concentrations of acids were much lower in the continuous-flow culture because of their continuous wash-out, the conversion rate of the acids into respective products was negligible. This led to the corresponding simplification of the model for the continuous-flow culture mode in the mass balances for both acids.

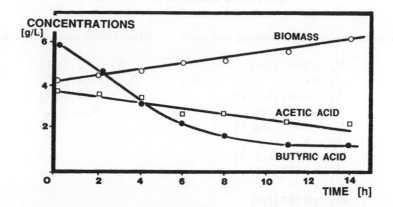

Figure 7.14: Summary of the experimental Continuous-Flow Throughput culture data (concentrations of biomass, butyric acid and acetic acid) for the acetone-butanol fermentation process, after Leung[15].

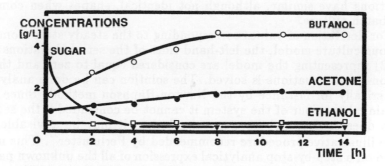

Figure 7.15: Summary of the experimental Continuous-Flow Throughput culture data (concentrations of the substrate and solvents) for the acetone-butanol fermentation process, after Leung[15].

TABLE 7.10

A-B-E Process Kinetic Model Parameters for
the Throughput Continuous-Flow Culture (Leung)[15]

Kinetic constants (parameters) and units: Other kinetic constants(parameters):

$k'_1 = 0.44$ (1/h)	$K_I = 6.0$ (g/L)
$k_2 = \star$	$K_{IBA} = 0.97$ (g/L)
$k_3 = 0.005$ (L/g· h)	$K_S = 1.0$ (g/L)
$k_4 = 1.25$ (1/h)	K_{AA}, K_{BA} $\star$
$k'_5 = 0.6$ (L/g· h)	
$k_6 = \star$	
$k'_7 = 0.367$ (L/g· h)	
$k_8 = 0.225$ (1/h)	
$k_9 = \star$	
$k_{10} = 0.187$ (1/h)	
$k_{11} = 0.033$ (1/h)	
$k_{12} = \dagger$	
$k_{13} = \dagger$	

$\star$ not possible to estimate for insufficient data. Zero (0) value assumed, except for k_2 which was taken as for the batch culture (Section 7.2.1).

$\dagger$ amount and composition of fermentation gases not determined.

b) *Computer process simulation and discussion*

Figure 7.16(a-g) presents the summary of specific rate relationships for individual basic biosynthetic steps of the A-B-E process as considered for the construction of the continuous-flow relationships. It is seen that these functions have similar, although not identical, shapes when compared to the batch culture.

For derivation of values corresponding to the steady state from the continuous culture model, the left-hand sides of the set of Equations (7.25) to (7.32) representing the model are considered equal to zero and the system of non-linear equations is solved. The solution can be done analytically or numerically for instance by the Newton-Raphson method. Since from the dynamic behaviour of the system it cannot be concluded if the set of equations does or does not have a real solution, it may be advisable to use a more illustrative procedure recommended by Perlmutter[22]. This method is based on a step-by-step analytical expression of all the unknown parameters in the butanol material balance equation, i.e., the concentrations of sugar, biomass, butyric acid, and dilution rate, as functions of butanol concentration. By doing so, one non-linear algebraic equation is obtained: $f(B) = \cdots$ whose value with regard to butanol concentration and initial sugar concentration can be "mapped" by using a computer.

An example of results of such "mapping" done for the Cell-Retention Continuous-Flow Culture (Section 7.2.2) are presented in Figures 7.26 and

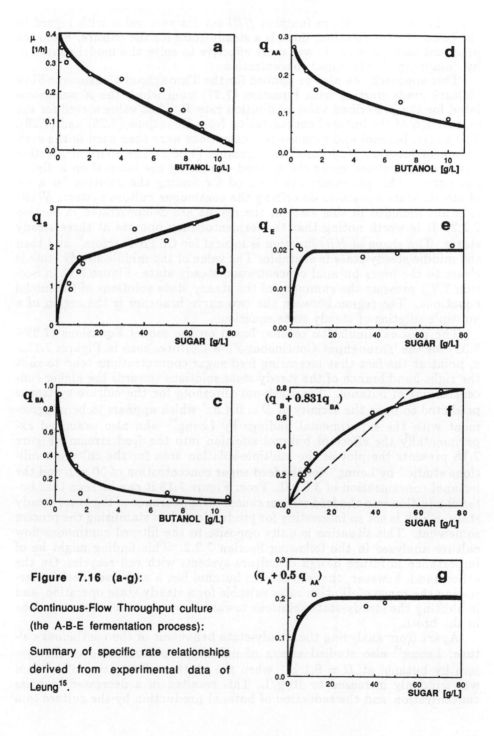

Figure 7.16 (a-g):

Continuous-Flow Throughput culture

(the A-B-E fermentation process):

Summary of specific rate relationships

derived from experimental data of

Leung[15].

7.27. In the points where function $f(B)$ attains zero value with regard to the butanol concentration, there is a steady-state for the culture. From the practical point of view, it was more effective to solve the model equations by "mapping" in the sugar concentration.

This approach can also be applied for the Throughput Continuous-Flow Culture mode starting with Equation (7.27) from which the X was calculated for the prescribed value of dilution rate D. That value served for the calculation of the butanol concentration from Equations (7.25) and (7.26). The sugar, butanol and biomass concentrations were then used in the evaluation of function $f(B)$ based on Equation (7.30). The roots of function $f(B)$ were calculated using the method of halving the interval on a digital computer. This procedure can be used for finding the solution for a set of steady state equations describing the continuous culture system. While it is not included in this section, the results are demonstrated in Section 7.2.2. It is worth noting that the fermentor may operate at three steady states. The shape of $f(B)$ function is typical for CSTR reactors[22] and then the middle steady state is unstable. The value of the middle steady state is closer to the lower butanol concentration steady state. Figure 7.17 in Section 7.2.2 presents the summary of the steady state solutions of the model equations. The region between the two curve branches is the region of a multiple solution of steady state equations.

The process simulation results, based on the model Equations (7.25 - 7.32) for the Throughput Continuous-Flow Cultures seen in Figures 7.17a-c, point at the fact that increasing feed sugar concentrations tend to shift the right-hand branch of the steady-state solutions towards the higher concentrations of butanol. The wash-out threshold for the culture system is predicted to be in the vicinity of $D = 0.4 \text{ h}^{-1}$ which appears to be in agreement with the experimental findings by Leung[15] who also examined experimentally the effect of butanol addition into the feed stream. Figure 7.18 presents the plot of the multiple-solution area for the culture conditions studied by Leung[15] for the feed sugar concentration of 50 g/L and the butanol concentration of 3.6 g/L. From Figure 7.18 it can be seen that butanol addition into the feed stream causes disappearance of the lower steady state, which is not so interesting for production, thus stabilizing the process somewhat. This situation is quite opposite to the filtered continuous-flow culture analyzed in the following Section 7.2.2. This finding might be of importance in future design of culture systems with cell recycle. On the other hand, however, the feed-stream butanol has a negative effect in narrowing the range of dilution rates suitable for a steady state operation, and in shifting the steady-state solutions towards lower butanol concentrations in the broth.

Apart from analyzing the steady-state behaviour of the continuous culture, Leung[15] also studied effects of pulse-type perturbation of the system by butanol at $D = 0.1 \text{ h}^{-1}$ when the broth concentration of butanol was suddenly increased to 16 g/L. This resulted in a decreased biomass concentration and the reduction of butanol production by the culture to a

Figure 7.17 (a-c):

Continuous-Flow Throughput culture
(the A-B-E fermentation process):

Process simulation results for
three different values of
S_O = 50; 75 and 100 g/L.

The washout of the culture is predicted
in the vicinity of D = 0.4 h^{-1}.

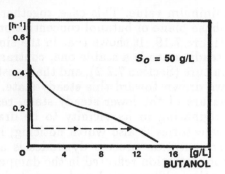

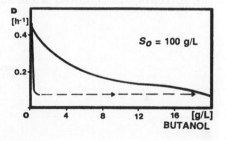

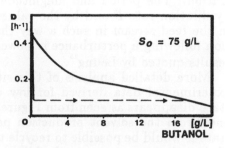

Figure 7.18:

Continuous-Flow Throughput culture
(the A-B-E fermentation process):
Process simulation results for S_O = 50 g/L
and butanol added to the feed stream at
B_O = 3.6 g/L.

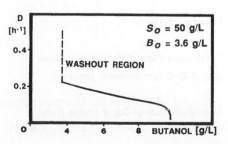

minimum value. This case, together with other possible trajectories in the phase plane of butanol concentration vs sugar concentration, is depicted in Figure 7.19. It shows that in the simple continuous-flow culture the upper steady state is a stable one, contrary to the cell-retention continuous-flow culture (Section 7.2.2), and that most of the trajectories (in the phase plane) are drawn toward this steady state. The simulation indicates an unstable nature of the lower steady state, resulting in the phase space trajectories originating in its vicinity to be drawn to the upper one. The shape of trajectories is also worth noticing, indicating that the stabilization of the fermentation process dynamics is affected by the inertia in the biomass concentration reflected in the dampened oscillations in the butanol concentration. The period and magnitude of these oscillations may be affected by dilution rate D and by the values of sugar and butanol concentrations in the feed stream in such a way that at a lower D the process stabilization following a perturbation is slower, in agreement with the experimental results quoted by Leung[15].

More detailed analysis of this situation would require more extensive experimental data derived for low dilution rates when the lower steady state disappears as seen from Figure 7.17a-c. In order to remove the lower and, from the solvent production point of view, not so attractive steady state it would be possible to recycle the supernatant broth from which part of the solvents was removed while sugar substrate was replenished.

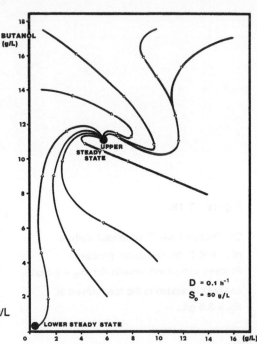

Figure 7.19:

Continuous-Flow Throughput culture (the A-B-E fermentation process):

Simulated Butanol-Sugar phase plane trajectories for $D = 0.1$ h^{-1} and $S_O = 50$ g/L show a stable "upper" steady state.

7.2.2 THE CELL-RETENTION CONTINUOUS-FLOW CULTURE SYSTEM

a) *The model development*

The cell-retention system is similar to the flow-through system in that the liquid phase is always being fed into and withdrawn from the bioreactor. In the cell retention operating mode, however, the biomass is retained in the bioreactor, increasing thus the "biocatalyst" concentration in the system. By either recycling the cells or selectively preventing them from leaving the bioreactor working volume, an increased volumetric productivity of the bioreactor can be achieved.

The A-B-E batch process kinetics model can be extended, as was similarly shown in the previous chapter, to also cover the conditions of the continuous-flow culture with retention of cells in the bioreactor (filtered system). The system dynamics at a constant volume V of fermentation broth and dilution rate $D(F/V)$ can be described by using the following differential equations considering the flow of liquid medium and accumulation of biomass.

Formally, the model for the cell-retention operating mode is similar to that developed for the flow-through system with the exception for the biomass mass balance which will assume the following form with the biomass withdrawal term (DX) omitted.

Growth:

$$\frac{dX}{dt} = 0.56(y - 1)\, X - k_2 XB \tag{7.33}$$

$$\frac{dy}{dt} = [\mu(S, B) - 0.56(y - 1)]\, y \tag{7.34}$$

$$\mu(S, B) = k_1 \frac{S}{S + K_s} \frac{K_I}{K_I + B} \tag{7.35}$$

Sugar material balance with sugar inlet concentration s_o:

$$\frac{dS}{dt} = -k_3 SX - k_4 \frac{S}{S + K_s} X - D(S_o - S) \tag{7.36}$$

Material balance for butyric acid:

$$\frac{d(BA)}{dt} = k_5 S \frac{K_I}{K_I + B} X - k_6 \frac{BA}{BA + K_{BA}} X - D \cdot BA \tag{7.37}$$

Material balance for acetic acid:

$$\frac{d(AA)}{dt} = k_8 \frac{S}{S + K_s} \frac{K_I}{K_I + B} X - k_9 \frac{AA}{AA + k_{AA}} X - D \cdot AA \quad (7.38)$$

Material balance for butanol:

$$\frac{dB}{dt} = k_7 S X - 0.831\left[\frac{d(BA)}{dt} + D \cdot BA\right] - D \cdot B \quad (7.39)$$

Material balance for acetone:

$$\frac{dA}{dt} = k_{10} \frac{S}{S + K_s} X - 0.5\left[\frac{d(AA)}{dt} + D \cdot A\right] - D \cdot A \quad (7.40)$$

Material balance for ethanol:

$$\frac{dE}{dt} = k_{11} \frac{S}{S + K_s} X - D \cdot E \quad (7.41)$$

The weight production rate of H_2 and CO_2 related to the broth volume:

$$\frac{dCO_2}{dt} = k_{12} \frac{S}{S + K_s} X \quad (7.42)$$

$$\frac{dH_2}{dt} = k_{13} \frac{S}{S + K_s} X \quad (7.43)$$

This set of equations represents the general model for the continuous-flow fermentation process with cell retention.

The model coefficients have been expressed ("identified") by fitting the model to original experimental data produced by Mulchandani[19,20]. Values of model coefficients are summarized in Table 7.11 together with calculated values for their respective Absolute and Relative Sensitivities.

It is essential to bear in mind that these coefficients have been expressed for experiments with a different strain of C. acetobutylicum (Leung[15]) and different culture conditions. This may, in part, be the reason for some differences when these coefficients are compared to those derived from the batch culture data (Table 7.7).

Upon comparison of the sensitivity analyses for the batch and the cell-retention continuous-flow culture models (Table 7.7 and 7.11) it can be seen that the most conspicuous changes are in the Relative Sensitivities of coefficients k_1, k_3, k_5, k_7, and K_I. The difference between the two culture systems becomes even more apparent when the coefficients for the two models are ranked in the sequence of their importance, which for the cell-retention continuous-flow culture model is as follows: k_4, K_I, k_5, K_S, k_7, and k_6. The altered sequence of the parameter importance indicates that in this culture mode the non-linear character of the process kinetics is more pronounced, particularly due to the effect of process inhibition by the product and also its limitation by the substrate. For the process control under these conditions the sugar material balance will play the most important role together with balances for the butyric acid and butanol. The eventual process control strategy will have to respect the system non-linearity with respect to eventual continuous model identification based on monitoring of the system.

The advantage derived from the sensitivity analysis of the model with regard to the parameters is in finding that the process control strategy does not need to be dependent on the direct measurement of the biomass concentration since its value can be calculated from the kinetic equations and experimental analytical data from chromatographic or other ways of determining the substrate concentration and those of the fermentation products. The comparison of process simulation results with experimental data for the transition phase between the batch and the cell-retention continuous-flow culture[19] is depicted in Figure 7.20. The analysis of experimental data from the later part of the experiment, when a half-hour interruption of the culture process resulted in a deep sugar limitation, showed that the kinetic behaviour of the system is governed by the same relationships. However, the numerical value of k_6 changed from 0.5 to 0.0 reflected in a slowdown of butanol biosynthesis and accumulation of butyric acid. As can be seen from the data published by Monot *et al.*[17], the glucose concentration in culture medium plays an important role during further conversion of acetic and butyric acids. If the sugar concentration is smaller than 5 g/L the conversion-utilization of the acids is affected resulting in lowering of the overall yield of neutral solvents.

The cell-retention continuous-flow system is a special type of a cultivation whereby the biomass is retained in the bioreactor while the products are freely leaving the system in a supernatant liquid. A real opposite is the throughput continuous-flow system whereby whole fermentation broth is withdrawn resulting thus in the minimum average biomass age in the system.

Basically, it is possible to conclude that the kinetics and values of model parameters established for different types of cultivations, with the exception of the inhibition constant K_I, are almost identical within the ranges of experimental errors. The inhibition mechanism may thus correspond to the physiological state of the culture represented by its age or cell wall composition, and is more pronounced with the time of exposure of the culture to the metabolic products. This working hypothesis, however, would require more thorough and certainly very experimentally demanding verification using age-segregated population models[24]. Solution of these models could

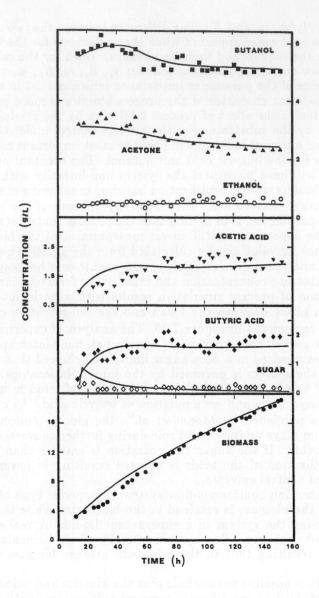

Figure 7.20: Continuous-Flow Cell-Retention culture (the A-B-E fermentation process): Comparison of the process computer simulations (solid line) and the experimental data, after Mulchandani and Volesky[20].

D = 0.089 h-1 S_O = 35 g/L   14.5 h < t < 53 h

S_O = 31 g/L t > 53

contribute in allowing a more general view of these cultivations, however, the methodology of this approach would be somewhat complex. For this reason, further considerations will be based on the model developed so far, assuming that the average culture age exceeds 10 h and the inhibition constant K_I is in the range of (0.43-0.83)g butanol/L.

b) *Computer process simulation and discussion*

The results of the model identification routine based on the experimental data by Mulchandani[19,20] are summarized in Table 7.11 which summarizes the values of model coefficients and their respective absolute and relative parametric sensitivities.

TABLE 7.11

Values of Kinetic Parameters and their Absolute and Relative Parametric Sensitivities for the Cell-Retention Continuous-Flow Culture at $D=0.113$ hr^{-1}

Parameter	Value	Units	Absolute sensitivity	Relative sensitivity
k_1	0.38	(L/g h)	8.138	0.223
k_2	0.0014	(L/g h)	-459.3	0.046
k_3	0.03	(L/g h)	-80.4	0.174
k_4	0.85	(1/h)	-26.9	1.645
k_5	0.270	(L/g h)	-36.3	0.706
k_6	0.050	(1/h)	-140.1	0.505
k_7	0.090	(L/g h)	-22.12	0.558
k_8	0.660	(1/h)	-1.61	0.076
k_9	0.01	(1/h)	0.28.81	0.027
k_{10}	0.132	(1/h)	-47.41	0.222
k_{11}	0.019	(1/h)	-5.9	0.006
k_{12}	0.6	(1/h)	-	-
k_{13}	0.019	(1/h)	-	-
K_S	1.0	(g/L)	9.54	0.688
K_I	0.42	(g/L)	24.2	0.733
K_{AA}	0.5	(g/L)	-0.105	0.004
K_{BA}	0.5	(g/L)	1.422	0.051

The identified model of the process was used for a computer simulation study of the fermentor behaviour at different dilution rates D and different sugar concentrations S_o. The implicit assumption was made concerning the medium composition which would not allow the culture limitation by nitrogen, and also that the culture growth can only be limited by sugar or inhibited by an increased butanol concentration. The batch process parameters were taken as initial conditions for the dynamic simulation.

TABLE 7.12

The Cell-Retention Continuous-Flow System
Initial Conditions for Dynamic Computer Process Simulation

$$
\begin{aligned}
S &= 3.3 \text{ g/L} & E &= 0.24 \text{ g/L} \\
X &= 4.24 \text{ g/L} & BA &= 5.88 \text{ g/L} \\
B &= 1.84 \text{ g/L} & AA &= 3.27 \text{ g/L} \\
A &= 0.81 \text{ g/L} & y &= 1.2
\end{aligned}
$$

The simulation was carried out for the values of $D = 0.025$; 0.05; 0.075; 0.1; 0.125 h^{-1}, and inlet sugar concentrations from 10 g/L to 100 g/L in 10 g/L increments. The simulation results are presented in Figures 7.21 – 7.25. The simulation was based on an adapted computer program derived from previously developed programs for model identification. The listing of this computer program is in Table 7.13 (Appendix).

From Figures 7.21–7.25 it can be seen that for the inlet sugar concentration lower than 40 g/L the process is self-stabilized within the range of sugar limitation. Sugar consumption is in equilibrium with growth and end-product formation. Acetic acid and butyric acid is gradually washed out of the fermentor because of the culture inhibition by butanol, or it is stabilized at a certain level when the butanol productivity is low. The biomass concentration exhibits a stabilizing characteristic with time.

Increasing the inlet sugar concentration above 40 g/L induces destabilization of the process, leading to increased production of butanol which in turn results in increased growth inhibition and decay of living microbial cells. The process enters a non-steady state operation which is likely to demonstrate itself by an oscillatory behaviour. The activity of microbial population is lowered until the inhibitory product (butanol) is partially washed from the fermentor by the incoming fresh medium. The model in its current form, however, is not capable of predicting the type of process instability. Further modifications and improvement of the model would be necessary which would likely lead to more complex forms of a structured model reflecting better the culture physiology.

It is of interest to analyze the modelled process situation in the area of high productivity of neutral solvents. As seen from the dynamic properties of the process model, the cell-retention continuous-flow culture system is affected by the inlet sugar concentration. Its increase promotes the butanol end-product formation which through a negative feedback mechanism has an effect on the process dynamics. The steady-state value in this culture system can be obtained by numerically solving the set of non-linear equations originating from the set of differential Equations (7.33)–(7.43) whereby their left sides are equal to zero.

The technique of "mapping" was used here as recommended by Perlmutter[22] and discussed briefly in previous Section 7.2.1. The results of such "mapping" using again the "butanol function" $f(B)$ are graphically summarized in Figure 7.26 with the feed sugar concentration S_o as a parameter.

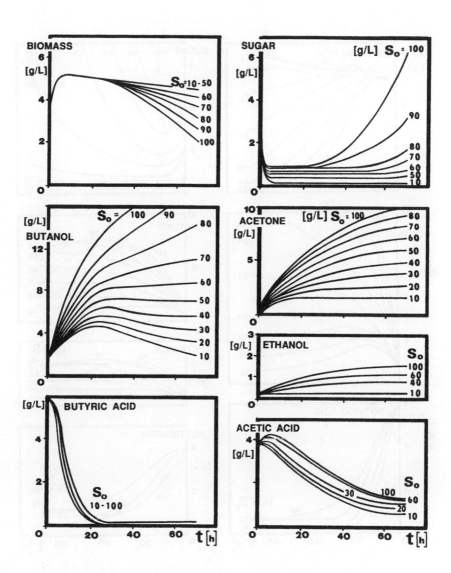

Figure 7.21: Continuous-Flow Cell-Retention culture (the A-B-E fermentation process): Computer simulation of the process dynamics for $D = 0.025$ h^{-1}.

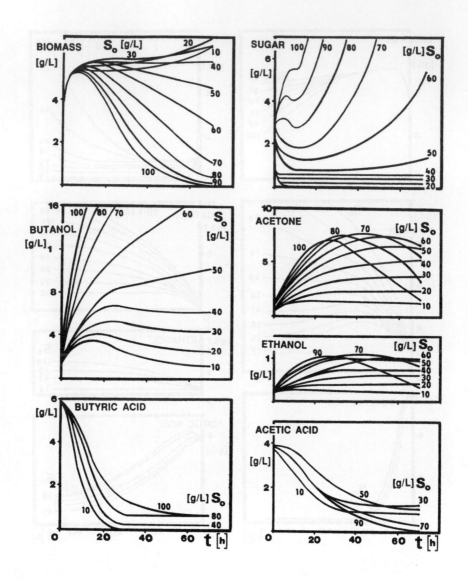

Figure 7.22: Continuous-Flow Cell-Retention culture (the A-B-E fermentation process): Computer simulation of the process dynamics for $D = 0.05$ h^{-1}.

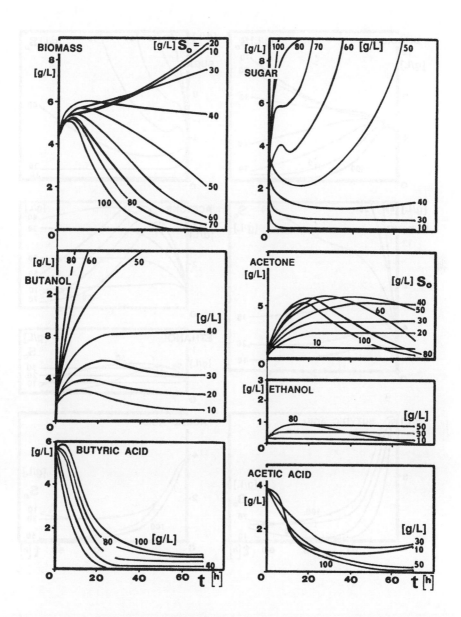

Figure 7.23: Continuous-Flow Cell-Retention culture (the A-B-E fermentation process): Computer simulation of the process dynamics for $D = 0.075$ h^{-1}.

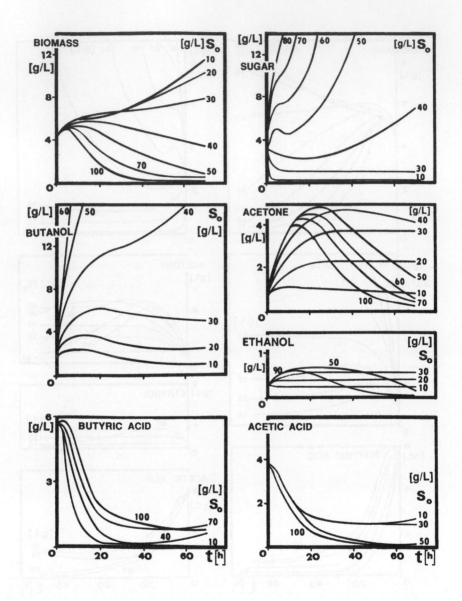

Figure 7.24: Continuous-Flow Cell-Retention culture (the A-B-E fermentation process): Computer simulation of the process dynamics for $D = 0.1$ h^{-1}.

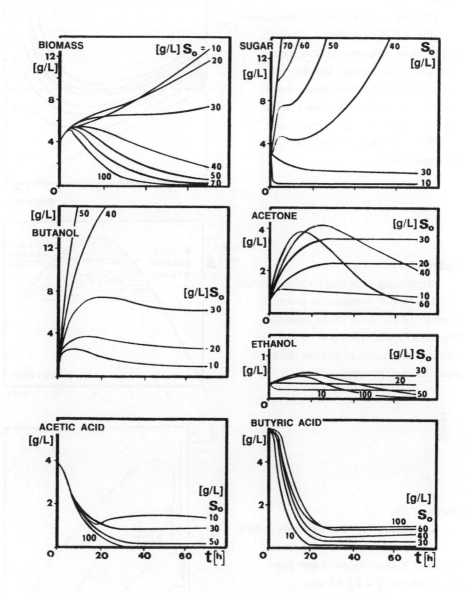

Figure 7.25: Continuous-Flow Cell-Retention culture (the A-B-E fermentation process): Computer simulation of the process dynamics for $D = 0.125 \ h^{-1}$.

Figure 7.26:

Continuous-Flow Cell-Retention culture (the A-B-E fermentation process): Steady state solutions of the model equations. The butanol function $f(B)$ synthesized from the mass balance model equations indicates the steady state where it passes through its zero value line. The feed sugar concentration (S_O) is the parameter.

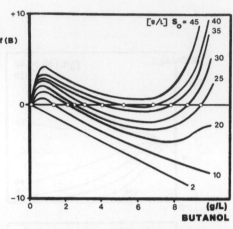

Figure 7.27:

Continuous-Flow Cell-Retention culture (the A-B-E fermentation process): Summary of the steady model equations. The area under the curve branches indicates the region of multiple steady states for the bioreactor..

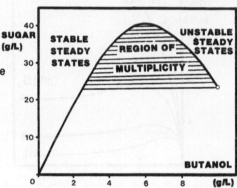

Figure 7.28:

Continuous-Flow Cell-Retention culture (the A-B-E fermentation process): Simulated Butanol-Sugar phase plane trajectories for $D = 0.1$ h-1 and $S_O = 30$ g/L show a stable "lower" steady state and an unstable "upper" steady state in the bioreactor behavior under the given operating regime.

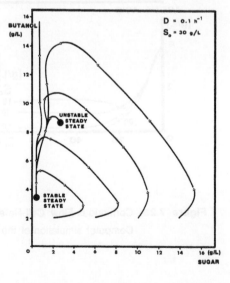

In the points where the "butanol function" $f(B)$ attains zero value with regard to the butanol concentration, there is a steady state for the culture. For low inlet sugar concentrations which bring the process operation into the region of deep substrate limitation, there is virtually no butanol formation. With increasing inlet sugar concentration, a steady state butanol concentration increases until the inlet sugar reaches 22 g/L and then function $f(B)$ goes through the zero line twice. It means that in this interval there can be two steady states, one which exists when the process operates under sugar limitation and the second one when it is inhibited by the end-product. While the first steady state is characterized by a higher biomass concentration and a lower butanol concentration, the second one is attained at higher butanol and lower biomass concentrations.

Increasing further the inlet sugar concentration leads to the two steady states drawing closer together, eventually becoming one at 40 g/L of sugar in the feed. Increasing S_o beyond this value results in function $f(B)$ having no solution for real butanol concentrations which means that there is no steady state for the cell-retention continuous culture process in this range of operation. In the simulation calculations based on the current rather simplistic model this may be indicated as an unrealistic model prediction of a butanol "explosion" and a loss of viability of the microbial population poisoned by the end-product.

Figure 7.27 presents the relationship between steady state butanol concentrations and the inlet sugar concentration. The marked area under the curve represents the area of multiple steady states. The Figure indicates that the highest butanol (and thus total solvents) productivity can be obtained if the process were conducted on the second branch of the curve where the steady state is controlled through butanol inhibition. Such operation will be characterized by better bioreactor performance as judged by the butanol productivity, however, residual sugar may be found in the broth, potentially complicating the solvent recovery procedure, decreasing the overall conversion yield and negatively affecting the process economy.

Prediction of the area where two steady states exist is a new qualitative step in understanding the potential of the cell-retention continuous-flow culture technique. The steady states in the flow-through bioreactor may be stable or unstable. For solution of the reactor stability problems, the whole mathematical apparatus could be used which was described by Perlmutter[22]. From the practical standpoint, however, the technique of the process simulation for different initial conditions can be used. This is how trajectories in the phase space between the two steady states can be obtained.

Figure 7.28 depicts the curves for the broth sugar and butanol concentrations at $D = 0.1$ h^{-1} and the inlet sugar concentration of 30 g/L. The lower steady state on the left-hand branch of the curve from Figure 7.26 is obviously stable while the upper steady state on the other branch is not. Interesting is the shape of trajectories (curves) which, when initiated from the higher butanol and sugar concentrations, have to by-pass the labile steady state to converge onto the lower stable steady state when the lower sugar concentrations are reached. This characteristic of the system could be used for controlling it at a high butanol productivity in such a way that the whole process would start in a batch mode from 50 g/L sugar and subsequently, when sugar concentration in the broth drops to 3 g/L or less, the continuous liquid feed and withdrawal would commence with a one-shot sugar addition to bring its concentration into a 5 - 10 g/L range. This would bring the process back into a range from which the labile steady state would have to be "by-passed" again. By using this strategy based on close monitoring of the process, it could be maintained in the region with high butanol and low acids concentrations in the broth. This may be possible particularly considering the relatively long time constants of the process.

When the cell-retention continuous-flow process is left without appropriate close control, it would have a tendency to assume the "lower" stable steady state characterized by a higher biomass concentration of solvents.

c) *Stabilizing the CF-CR culture process by the "bleed"*

The throughput continuous-flow system (Section 7.2.1), however, exhibited a high-productivity stable steady state whereby the lower-productivity steady state with a high biomass concentration practically could not be reached because of the dynamic characteristics of the system. The cell-retention culture high-productivity steady state could be maintained in such a way that a part of the whole culture broth can be withdrawn as a non-filtered whole broth stream. The mathematical model describing this situation can be derived through modifying the model from Section 7.2.1 in the following way:

The growth mass balance for a dynamic system will differ from Equation (7.25) in its convection part:

$$\frac{dX}{dt} = 0.56 \, (y - 1) \, X - k_2 \, BX - D_2 X \qquad (7.25')$$

The other mass balances will be identical to those in this Section [Equations (7.34)–(7.43)]. The dilution rate, however, has to be redefined, consisting of the "clear liquid" (D_{CR}) and the "whole broth bleed" (D_2) parts:

$$D = D_{CR} + D_2 \qquad (7.44)$$

Similarly as in the case of the full cell-retention culture system, the analysis of steady states will be done first. For $D_2 = 0$ the system becomes the full cell-retention culture one. When a whole broth harvest stream (bleed) including biomass is withdrawn from the system, the high productivity steady state can be established in the system at relatively low values of D_2. The culture system can then operate with three steady states typical for the throughput continuous-flow culture process. As can be derived from Figure 7.29, the high-productivity steady state is stable even with higher sugar concentrations in the feed stream and there is no butanol "instability" in

Figure 7.29:

Continuous-Flow Cell-Retention culture (the A-B-E fermentation process): Simulation of the whole culture broth "bleed" (D_2) effect on the stability of the reactor performance. Three multiple steady states can exist when D_2 is operated in addition to the "clear" liquid (no cells) dilution rate D.

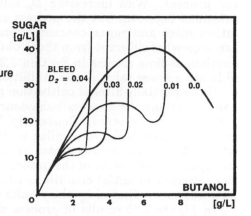

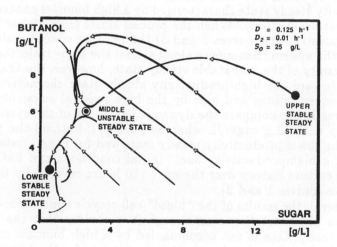

Figure 7.30: Continuous-Flow Cell-Retention culture (the A-B-E fermentation process): Simulated Sugar-Butanol phase trajectories for D= 0.125 h-1 and S_O = 25 g/L show an unstable "middle" steady state with the bioreactor performance stabilizing at either "lower" or "higher" steady states. Stabilizing effect of the "bleed" D_2 dilution rate on the bioreactor system is seen in comparison with Figure 7.28.

the process. With increasing D_2 value the area of occurence of multiple solutions decreases but the process can be carried out even with high dilution rates and sugar concentrations. Results presented in Figure 7.29 are somewhat different from those obtained from analyzing the throughput continuous-flow culture in Section 7.2.1 because the experimental data[19,20] indicate a lower value of the inhibition constant with regard to butanol.

If the partially-filtered cultivation process is operated at lower feed sugar concentrations the process behaviour similar to that presented in Figure 7.29 could be expected. Conversely, high feed sugar concentrations result in the process dynamics similar to that seen in Figure 7.19. In the case where three stable steady states exist, e.g. for $D_2 = 0.01$ and $S_o = 22$ g/L, the dynamic behaviour of the process can be expected which will depend on the selection of initial conditions because both extreme steady states will be stable, as indicated by the results derived earlier.

In Figure 7.30 results of process simulations are presented for different initial conditions in a state space of S, B coordinates. The Figure shows steady states as well as appropriate trajectories. When the process starts from lower butanol and sugar concentrations ($S \leq 10$ g/L), its behaviour is similar to that of the cell-recycle culture shown in Figure 7.28. First, the butanol concentration increases in the direction toward the middle non-stable steady state, and then the system tends to move toward the lower-productivity steady state characterized by a high biomass concentration but low solvent productivity. When the process starts from higher butanol and sugar concentrations (curves 2 and 3) the product concentration increases first and the system then moves along almost the same trajectory as before. In the vicinity of the non-stable steady state, however, the trajectories are diverted toward the high-productivity steady state characterized by high solvent concentrations and also by the high residual sugar concentration. It is interesting to compare the dynamic behaviour of the system following a start up with 10 g sugar/L whereby for 3 g/L butanol the process goes toward the lower-productivity steady state and for 6 g/L butanol it tends toward the high-productivity one. Initial conditions can basically determine the culture history over the next 150 hours covered by the computer simulation (curves 1 and 2).

In general, the results of the "bleed" cell-recycle continuous-flow culture simulation are interesting even though it is obvious that the solvent production steady state is not accompanied by a high biomass concentration and it is likely to be very close to the steady state for the conventional continuous flow culture without the filter (cell retention). An important point is the prediction of possible harvesting flow rates in the interval between 0 to 0.01 h^{-1}. As indicated by the simulation results, the "bleed" cell-retention culture system would not result in higher solvent productivities when compared to the continuous-flow or even to the batch culture systems. Understanding of the dynamic behaviour of this culture system, however, is to be of advantage in explaining certain process phenomena which are likely to occur in more complex experimental culture arrangements.

7.3 THE FED-BATCH CULTURE SYSTEM

a) *The model development*

The fed-batch culture system is a dynamic non-steady state culture system whereby the culture broth volume is steadily increasing because of the fresh medium addition. There is no fermentation broth withdrawal during the course of the fermentation. In fermentation technology this is a commonly used approach for extending the productive phase of a fermentation process characterized by a low growth rate with microorganisms converting added substrate into a desired product at the highest yield. While deriving the material balances describing the dynamics of the fed-batch culture, the volumetric accumulation of broth in the bioreactor has to be also taken into account, expressed in a general form as:

$$\frac{dV}{dt} = F'$$

(7.45)

where F' is a feed flow rate expressed in the same volumetric V units as the batch culture. The feed flow rate may be either constant with a fixed feed pump adjustment or variable, based on the process requirements. These requirements may be based on maintaining the concentration of a certain component in the broth at a constant level or having it follow a certain pattern which has been established according to a certain strategy of process optimalization. The fed-batch culture system offers a great variety of options with regard to the optimal product accumulation thus maximizing productivity of the bioreactor.

This section deals with problems related to the design of the fed-batch fermentation system using a mathematical process model for the computer simulation of the bioreactor behaviour. Based on the kinetics of microbial growth and product formation in a batch system as dealt with in Section 7.1, the differential material balances can be written for the fed-batch culture system using the concept of the "marker" of the physiological state of the culture.

Material balance for the biomass:

$$\frac{dX}{dt} = 0.56(y - 1)X - k_2 BX - \frac{1}{V}\frac{dV}{dt}X$$

(7.46)

$$\frac{dy}{dt} = [k_1 S \frac{K_I}{K_I + B} - 0.56(y - 1)]\, y$$

(7.47)

Material balance for the sugar substrate with feed concentration S_o:

$$\frac{dS}{dt} = -k_3 S X - k_4 \frac{S}{S + K_s} X + \frac{1}{V}\frac{dV}{dt}(S_o - S) \tag{7.48}$$

Material balance for butyric acid:

$$\frac{d(BA)}{dt} = k_5 S \frac{K_I}{K_I + B} X - k_6 \frac{BA}{BA + K_{BA}} X - \frac{1}{V}\frac{dV}{dt} BA \tag{7.49}$$

Material balance for acetic acid:

$$\frac{d(AA)}{dt} = k_8 \frac{S}{S + K_s}\frac{K_I}{K_I + B} X - k_9 \frac{AA}{AA + K_{AA}} X - \frac{1}{V}\frac{dV}{dt} AA \tag{7.50}$$

Material balance for butanol:

$$\frac{dB}{dt} = k_7 S X - 0.831\left[\frac{d(BA)}{dt} + \frac{1}{V}\frac{dV}{dt} BA\right] - \frac{1}{V}\frac{dV}{dt} B \tag{7.51}$$

Material balance for acetone:

$$\frac{dA}{dt} = k_{10}\frac{S}{S + K_s} X - 0.5\left[\frac{d(AA)}{dt} + \frac{1}{V}\frac{dV}{dt} AA\right] - \frac{1}{V}\frac{dV}{dt} A \tag{7.52}$$

Material balance for ethanol:

$$\frac{dE}{dt} = k_{11}\frac{S}{S + K_s} X - \frac{1}{V}\frac{dV}{dt} E \tag{7.53}$$

Material balances for the gases:

$$\frac{dCO_2}{dt} = k_{12}\frac{S}{S + K_s} V X \tag{7.54}$$

$$\frac{dH_2}{dt} = k_{13}\frac{S}{S + K_s} V X \tag{7.55}$$

Equations (7.46) to (7.55) represent the mathematical model of the fed-batch culture system.

b) *Process computer simulation and discussion*

The simplest technique for the fed-batch cultivation is based on a continuous feed stream of fixed-composition culture medium being metered into a bioreactor by a pump at a constant flow rate. This arrangement leads to a gradual increase of the bioreactor broth volume resulting in dilution of already synthesized products. Figure 7.31 presents the results of the fed-batch process computer simulation for the feed rate of 0.05 feed-stream volume per initial batch volume and hour. The initial bioreactor broth volume is doubled in 20 hours. The start of the feeding was simulated as a time event after 10 hours of initial batch cultivation of *Clostridium acetobutylicum*, for different feed sugar concentrations. Figure 7.31 shows that an arrangement considering constant feeding of the A-B-E process would not lead to an overall increase in biomass concentration. However, the acetic acid accumulation is partially suppressed while the most pronounced effect of fed-batch system conditions will be apparent in increased butanol and acetone concentrations. The simulation results also indicate that increasing feed concentration of sugar results in residual sugar in the bioreactor which adversely affects the overall process economics.

For process optimization, it is useful to express even a simple overall function (f_{OPT}) to be optimized. This function can be expressed for example as a carbon balance-based sum of desirable process products (+) and the undesirable ones (–), including the residual substrate, at the end of the fed-batch run:

$$f_{OPT} = \frac{V_o}{V_f}(0.6486B + 0.622A + 0.5217E - 0.4S - 0.4806X - 0.4AA - 0.4865BA) \qquad (7.56)$$

where the value of V_o/V_f expresses the ratio between the starting batch volume V_o and the final (terminating) bioreactor broth volume V_f. It is necessary to use this factor when the comparison between the batch and fed-batch systems at different feed rates is made since it expresses the effective utilization of the bioreactor volume in terms of its productivity. When a portion of the bioreactor volume is empty during the fermentation, even at high feed rates, it represents an unused volume when compared with a standard batch operation. However, this obvious disadvantage may be partially offset by considering the medium preparation and sterilization time which, in the fed-batch operated system, may overlap with the actual cultivation (post-inoculation) time. An overall preparation–cultivation time analysis has to be then considered for comparison of the operating effectiveness for the batch and fed-batch processes.

As seen from the results of the simulations for the fed-batch process started at either 25 g/L or 50 g/L of glucose with feeding initiated at either hour 8 or hour 10, for a certain feed rate there is an optimum with regard to the feed glucose concentration. Figure 7.32a presents the plot of function f_{OPT}, defined by Equation (7.56), versus the feed flow rate F. Figure 7.32b is a relationship for the feed sugar concentration and the feed flow rate. The

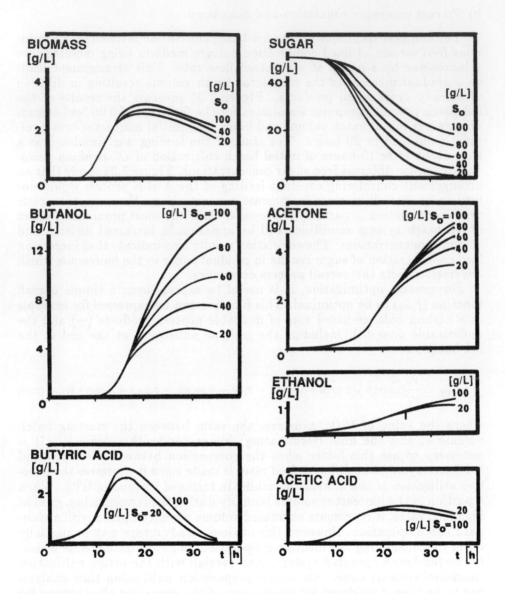

Figure 7.31: Fed-Batch Culture (the A-B-E fermentation process):

Computer simulation of the process dynamics for the volumetric feed
rate $F = 0.05$.

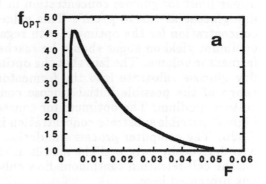

Figure 7.32: Fed-Batch Culture (the A-B-E fermentation process):

    a) The calculated profile of the f_{OPT} function for different volumetric feed rate F values;

    b) The relationship of the feed sugar concentration (S_O) and the volumetric feed rate F.

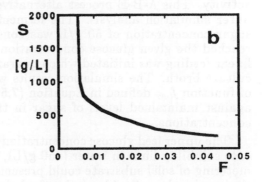

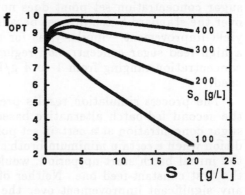

Figure 7.33:

    Fed-Batch Culture

(the A-B-E fermentation process):

    Calculated profiles of the relationship between f_{OPT} and the sugar concentration in the fermentation broth (S) for different feed sugar concentrations (S_O).

upper limit for glucose concentration in the feed is given by the density of dry glucose at 1.5416 g/cm^3. This value also determines the feed glucose concentration for the optimum with regard to the constant feed rate. The optimum yield on sugar should be reached when feeding 6.2 g glucose/h·L fermentor volume. The fact that the optimum is represented by feeding pure dry glucose substrate into the fermentor is generally valid for the whole range of the possible initial glucose concentrations in the starting batch culture medium. The optimum for constant feeding of glucose assumes the highest possible substrate concentration in the feed stream as seen in Figure 7.32b. The computer process simulation program for the fed-batch system was derived by only a slight modification of that presented in Table 7.12 for the cell-retention continuous-flow culture system. Its listing is therefore not presented here.

Another fed-batch process alternative is the one based on a constant sugar concentration in the feed stream because, for a number of microbial processes, it guarantees the biological stability of production. This approach is being extensively used in the production of antibiotics, bakers yeast, and amino acids. In these processes the effect of product concentration is not demonstrated by a negative feedback on the culture metabolic activity. This A-B-E process alternative was evaluated by using the computer simulation analysis. The fermentation process with starting batch sugar concentration of 50 g/L was considered. When the simulation run reached the given glucose concentration in the fermentation broth, a nonlinear feeding was initiated which guaranteed constant level of sugar in the culture broth. The simulation results were used for expressing the values of function f_{OPT} defined in Equation (7.56) which are plotted in Figure 7.33 against maintained levels of sugar in the broth for different feed glucose concentrations.

The upper feed glucose concentration of 400 g/L which is close to the glucose solubility limit in water (450 g/L), was considered here. Programmed metering of solid substrate could present some hardware problems thus justifying the above liquid solution feed choice.

The simulation results show that the feeding alternative with a broth sugar concentration set point does not present a significant improvement over the standard batch process because there is only an approximate 15% yield improvement. Similarly, as for the constant-feed fed-batch process, a high feed sugar concentration regime is preferred with the broth sugar concentration ranging from 1 to 4 g/L according to the feed sugar levels used.

The process simulation results presented in this section indicate that the second fed-batch alternative based on feeding to maintain the broth sugar concentration at a certain set point, considering a one-shot sugar addition when a certain minimum broth sugar concentration is reached during the initial batch start-up period, would result in a similar productivity to the first constant-feed one. Neither of these fed-batch systems represents any significant improvement over the conventional batch culture system unless perhaps an overall fermentor preparation-operation cycle is considered. Such consideration, however, is somewhat beyond the scope of this elaboration based on bioreactor kinetics.

Since the two above mentioned fed-batch alternatives, the constant feed flow and the constant broth sugar level maintenance, do not seem to represent an optimum fed-batch arrangement, the process simulation model will be further used to design the optimal feeding arrangement. In the area of control engineering there is a number of publications available in the literature dealing with a similar problem particularly concerning the chemical reactor performance optimization. It would be possible to use Pontryagin's maximum principle or methods of variable calculus for approaching this task but considering the strongly non-linear character of the model differential equation set, it would be necessary to use iterative numerical procedures for expressing the numerical values of the optimum profile. It may be satisfactory to attempt to find an approximate shape of the optimum feeding profile by using the software developed and applied in the previous sections of this text. The unknown profile can be approximated by a known function with parameters to be determined from the solution of the model equations and the optimum value of function f_{OPT} defined in Equation (7.56). The unknown profile can be approximated either in sections by a linear relationship or by the Taylor series for a finite time of a fermentation run when the feed rate has to be equal to zero.

$$F \;=\; F(t_F) \;+\; \frac{\mathrm{d}F(t_F)}{\mathrm{d}t}(t - t_F) \;+\; \frac{1}{2!}\frac{\mathrm{d}^2 F(t_F)}{\mathrm{d}t^2}\,(t - t_F)^2 \;+\cdots \qquad (7.57)$$

When the condition of the zero feed rate at the end of a fermentation run is used, then the optimal profile can be expressed as a polynomial of the following type:

$$F \;=\; a_1(t - t_F) \;+\; a_2(t - t_F)^2 \;+\; a_3(t - t_F)^3 \;+\; \cdots \qquad (7.58)$$

where unknown values of a_1, a_2, a_3, $\cdots$ are determined by a simulation as an optimum with regard to function f_{OPT} defined by Equation (7.56) and to the boundary condition for $F \geq 0$.

A program was assembled for the given kinetics which enables finding the maximum of function F with regard to parameters a_1, a_2, a_3, $\cdots$, which was based on the use of a subroutine for multivariable optimalization ROS and simulation programs ODE and RK4 in an arrangement similar to the one used for model identification outlined in Figure 7.12. The procedure of searching for an optimal feeding profile consisted of steps where the optimum was first found for one, then two, and eventually three members of the Taylor series in Equation (7.57). This approach is summarized in Table 7.14. The results of the fermentation process simulation with an optimal feeding profile are shown in Figure 7.34, indicating that for a good approximation of the optimal feeding profile a second order polynomial will be adequate.

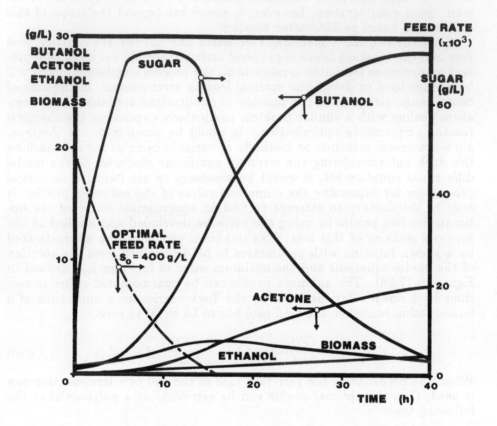

Figure 7.34: Fed-Batch Culture (the A-B-E fermentation process):

Simulated profiles of the key process variables during the fed-batch process carried out with the optimal feed rate profile. Note the maximum butanol concentration corresponding to the minimum sugar concentration left in the fermentation broth at the end of the fermentation run when the fermentor is full.

The optimum feed rate profile has been calculated for the highest feasible feed sugar concentration S_0 = 400 g/L. The feeding ceases when the fermentor is filled at hour 15 into the fermentation. The process is then carried out in a batch mode until termination at hour 40.

TABLE 7.14

Results of the Optimal Feed Rate Synthesis
for the Fed-Batch Culture System

Number of parameters	a_1	a_2	a_3	f_{OPT}
1	-3.359×10^{-4}	–	–	26.42
2	1.781×10^{-3}	7.352×10^{-5}	–	35.54
3	$1.79 \ \times 10^{-3}$	7.156×10^{-5}	-5.265×10^{-8}	35.57

Parameters of the simulation calculation:
$t_F = $ 40 h
$S_o = $ 400 g/L glucose in the feed stream

Initial concentrations at the feed time:
Biomass   0.6 g/L
Glucose   10 g/L
Butanol   0.5 g/L

The optimal feeding profile is characterized by a decreasing feeding rate up to hour 15 of the fermentation when the feeding ceases altogether and the process becomes a batch fermentation. The major contribution of the optimal feeding profile is in shortening the culture lag phase following the inoculation whereby the inoculum concentration shock is avoided. Because of the high initial feed rate the biomass growth rate continues unimpeded since the fermentation intermediate and end product accumulation is avoided minimizing thus their inhibition effect. The character of the optimal feeding rate differs from the constant or from the "sugar set point" feeding strategy to a considerable degree explaining their respective low effectiveness in improving the process performance.

Analysis of the fed-batch fermentation process by the process computer simulation led to evaluation of the effectiveness of different modes of feeding, resulting in establishment of the optimal feeding strategy. A low starting medium sugar concentration was indicated to be advantageous resulting in elimination of the culture lag phase following inoculation. The optimal feeding schedule recommends a high initial feeding rate of fresh medium containing a high glucose concentration (400 g/L) with an almost linear decrease in the pumping rate to the point when 60 – 70 g/L glucose concentration in the broth is reached. From this point, which is reached after approximately 15 hours, the bioreactor has a full volume and the fermentation is concluded in a batch mode until all the sugar is consumed. Neither the constant feeding nor the "constant sugar in the broth maintenance" feeding could be considered as viable process alternatives.

7.4 IMMOBILIZED CELLS CULTURE SYSTEMS

7.4.1 IMMOBILIZED-CELL SYSTEM IN A CSTR WITH A GROWTH-SUPPORTING MEDIUM

a) *The model development*

An immobilized-cell system with *Clostridium acetobutylicum* in gel phase beads is considered here. Growth and/or biochemical activity of the cells in the gel particle results in the establishment of concentration gradients, of sugar substrate and fermentation end-products throughout the bead, which in turn is reflected in the biomass concentration gradient established in the gel phase. At low biomass concentration (up to 10–15 g dry weight/L of gel) the gradients inside the gel particle are not that significant, however, when the biomass concentration inside the gel reaches high values, there is a slow-down of the fermentation process due to mass transfer limitations resulting from diffusion and interphase transfer limits. Since the reaction rates in a dispersed submerged culture system are known, an estimate of the process slow-down by transport phenomena can be made using the data published by Leung[15] for pseudo steady states where concentrations of 60–80 g biomass/L gel were reached. As a measure of the effect of diffusion, so called effectiveness factors can be employed which are widely used in dealing with the heterogeneous catalysis systems for expressing the ratio of the overall reaction rate related to the concentration in the particle and the rate for an ideally dispersed culture. The effectiveness factor for products can be derived from an overall mass balance as:

$$\eta_P = \frac{\dfrac{F'}{V_W}P - q_{PW}\,X_W}{\dfrac{V_G}{V_W}\,q_{PW}\,X_G} \tag{7.59}$$

For the fermentation substrate the effectiveness factor is:

$$\eta_S = \frac{\dfrac{F'}{V_W}(S_o - S) - q_{SW}\,X_W}{\dfrac{V_G}{V_W}\,q_{SW}\,X_G} \tag{7.60}$$

F' - overall feed rate

S_o - initial sugar concentration

V_w and V_G are volumes of water and gel phases in the reactor, respectively

X_w and X_G are biomass concentrations in water and gel phases, respectively

S -sugar concentration

P - product concentration

q_w is the specific rate of substrate consumption or production of a product for an ideally mixed culture.

The same approach to analyzing the batch and cell-retention continuous-flow culture data as presented in Sections 7.1 and 7.3 was also applied in evaluating the stirred immobilized-cell culture system results presented by Leung[15]. Figure 7.16a-g presents the specific rates of individual biosynthesis steps comprising the fermentation process. It is seen that these functions have a similar, although not identical, shape when compared to the batch culture data plots (Section 7.1). In order to be able to specify the difference(s) between the batch and the continuous-flow culture modes, the kinetic coefficients for this model were evaluated based on somewhat limited experimental data available. The results of the graphical data evaluation are presented in Table 7.10. Upon comparison of parameters which can be evaluated by this method, it can be seen that the batch kinetics differs from that of the continuous-flow culture in the value of the inhibition constant K_I which attains by an order of magnitude higher values ($K_I = 6$) for the continuous culture mode. The differences in the values of coefficients k_7, k_8, and k_{10} could be perhaps even attributed to inaccuracies in reading experimental data points from the literature used[15].

Results of the effectiveness factor evaluations for particle sizes 3.8 and 5 mm respectively, are presented in Table 7.15. It is seen that the transport phenomena-caused process slow-down concerning the substrate uptake and product formation is approximately one half except where butanol is the controlling component of the process, in which case this effect is more pronounced and the corresponding value is very high. This is probably caused by the fact that when the butanol concentration is high, the sensitivity of other metabolic systems is not variable with regard to that parameter. The behaviour of the culture system considered in this Section is different from the cell-recycle continuous-flow system where there is no space separation between the biomass and aqueous phase and the effectiveness factor is always equal to unity.

TABLE 7.15

Effectiveness Factors from the Analysis of
Experimental Data by Leung[15]

Component	η (d_p = 3.8 mm)	η (d_p = 5mm)
Sugar	0.41	0.45
Butanol	0.78	0.90
Acetone	0.30	0.23
Ethanol	0.53	0.41

Even though a greater part of the bioconversion takes place in the gel phase (80%), as derived from the process material balance, for the construction of the process model it is necessary to also consider the dynamic behaviour of the water phase. A certain amount of biomass leaks from the immobilizing gel phase and its transfer into the water phase is also assisted by escaping gases. From the experimental data it was established that the rate of biomass transfer is proportional to its concentration in the gel phase:

$$r_{XG} = k_2 X_G \qquad (7.61)$$

where $k_2 = 0.034$ L/g·h for both considered diameters of gel particles.

The mass balance of the system can be written in the form of the following relationships:

Material balance for biomass:

$$\frac{\partial X_G}{\partial t} = [\, 0.56\, (y_G - 1) - k_2 X_G \,]\, X_G \qquad (7.62)$$

$$\frac{\partial y_G}{\partial t} = [\, \mu\, (S_G,\ B_G) - 0.56\, (y_G - 1) \,]\, Y_G \qquad (7.63)$$

where

$$\mu\, (S_G,\ B_G) = k_1' \,\frac{S}{S + K_s}\, \frac{K_I}{K_I + S} \qquad (7.64)$$

Material balance for sugar substrate:

$$\frac{\partial S_G}{\partial t} = D\nabla^2 S_G - k_3 S_G X_G - k_4 \,\frac{S_G}{S_G + K_s}\, X_G \qquad (7.65)$$

where parameter ∇^2 for a spherical symmetry is:

$$\nabla^2 = \frac{\partial}{\partial r^2} + \frac{2}{r}\, \frac{\partial}{\partial r} \qquad (7.66)$$

Material balance for butanol in the gel phase:

$$\frac{\partial B_G}{\partial t} = D\nabla^2 B_G + \left[k_7' - 0.831\frac{k_5' B_G}{B_G + K_5} \right] \frac{S_G}{S_G + K_5} X_G \qquad (7.67)$$

Material balance for butyric acid in the gel phase:

$$\frac{\partial (BA)}{\partial t} = D\nabla^2 (BA)_G + k_5 \frac{B_G}{B_G + K_I} \frac{S_G}{S_G + K_s} X_G \qquad (7.68)$$

Material balance for acetone in the gel phase:

$$\frac{\partial A_G}{\partial t} = D\nabla^2 A_G + \left[k_{10} - 0.5k_8 \frac{K_I}{K_I + B_G} \right] \frac{S_G}{S_G + K_s} X_G \qquad (7.69)$$

Material balance for acetic acid in the gel phase:

$$\frac{\partial (AA)_G}{\partial t} = D\nabla^2 (AA)_G + k_8 \frac{K_I}{K_I + B_G} \frac{S_G}{S_G + K_s} X_G \qquad (7.70)$$

Material balance for ethanol in the gel phase:

$$\frac{\partial E_G}{\partial t} = D\nabla^2 E_G + k_{11} \frac{S_G}{S_G + K_s} X_G \qquad (7.71)$$

In order to complete the gel phase material balances the system of equations has to be supplemented by initial and boundary conditions. For simplicity, the start-up is assumed from a clean fermentor at time $t = 0$, when pellet diameter r changes from 0 to R, then there is:

$$S_G = (BA)_G = B_G = E_G = A_G = (AA)_G = 0$$

$$X_G = X_o \quad Y_G = 1$$

At time $t \geq 0$, the center of the pellet holds a symmetry condition:

$$r = 0 \qquad \frac{\partial S_G}{\partial r} = \frac{\partial B_G}{\partial r} = \frac{\partial (BA)_G}{\partial r} = \frac{\partial A_G}{\partial r} = \frac{\partial (AA)_G}{\partial r} = 0 \qquad (7.72)$$

For the outer surface of the pellet applies the condition for mass transfer when $r = R$:

$$D\frac{\partial S_G}{\partial r} = \beta(S_w - S_G) \tag{7.73}$$

$$D\frac{\partial B_G}{\partial r} = \beta(B_w - B_G) \tag{7.74}$$

$$D\frac{\partial (BA)_G}{\partial r} = \beta[(BA)_w - (BA)_G] \tag{7.75}$$

$$D\frac{\partial A_G}{\partial r} = \beta(A_w - A_G) \tag{7.76}$$

$$D\frac{\partial (AA)_G}{\partial r} = \beta[(AA)_w - AA_G] \tag{7.77}$$

$$D\frac{\partial E_G}{\partial r} = \beta(E_w - E_G) \tag{7.78}$$

Apart from the material balances in the gel phase, similar balances have to be written for the water phase which for the experimental arrangement considered can be modelled as an ideal mixer with volume V_w.

Material balance for biomass in the water phase:

$$V_w\frac{\mathrm{d}X_w}{\mathrm{d}t} = V_G X_G k_2 - 0.56 V_w(y_w - 1)X_w - F'X_w \tag{7.79}$$

$$\frac{\mathrm{d}y_w}{\mathrm{d}t} = [\,\mu\,(S_w, B_w) - 0.56\,(y_w - 1)\,]\,y_w \tag{7.80}$$

Material balance for sugar substrate in the water phase consists of sugar consumption in the water phase, transport of sugar into the gel phase and feeding of fresh substrate:

$$V_w\frac{\mathrm{d}S_w}{\mathrm{d}t} = -V_G a\beta(S_w - S_G) - V_w\left[k_4\frac{S_w}{S_w + K_s} + k_3 S_w\right]X_w +$$
$$+\, F'(S_o - S_w) \tag{7.81}$$

Material balance for butanol in the water phase will be similar:

$$V_W \frac{\mathrm{d}B_W}{\mathrm{d}t} = -V_G a\beta(B_W - B_G) +$$

$$+ V_W \left[k_7' - 0.831 k_5 \frac{B_W}{K_I + B_W} \right] \frac{S_W}{S_W + K_s} X_W - F'B_W \qquad (7.82)$$

Material balance for butyric acid in the water phase:

$$V_W \frac{\mathrm{d}(BA)_W}{\mathrm{d}t} = -V_G a\beta[(BA)_W - (BA)_G] +$$

$$+ V_W k_5 \frac{B_W}{B_W + K_s} \frac{S_W}{S_W + K_s} X_W - F'(BA)_W \qquad (7.83)$$

Material balance for acetone in the water phase:

$$V_W \frac{\mathrm{d}A_W}{\mathrm{d}t} = -V_G a\beta(A_W - A_G) +$$

$$+ V_W \left[k_{10} - 0.5 k_8 \frac{K_I}{K_I + B_W} \right] \frac{S_W}{S_W + K_s} X_W - F'A_W \qquad (7.84)$$

Material balance for acetic acid in the water phase:

$$V_W \frac{\mathrm{d}(AA)_W}{\mathrm{d}t} = -V_G a\beta[(AA)_W - (AA)_G] +$$

$$+ V_W k_8 \frac{K_I}{K_I + B_W} \frac{S_W}{S_W + K_s} X_W - F'(AA)_W \qquad (7.85)$$

Material balance for ethanol in the water phase:

$$V_W \frac{\mathrm{d}E_W}{\mathrm{d}t} = -V_G a\beta(E_W - E_G) + V_W k_{11} \frac{S_W}{S_W + K_s} X_W - F'E_W \qquad (7.86)$$

With the exception of k_2, values of rate parameters have been identified for the continuous-flow throughput culture mode (Section 7.2.1) and are summarized in Table 7.10. The mass transfer coefficient for diffusion in pure gel is quoted in Leung's dissertation[15] as 1.6×10^{-5} cm^2/s for solvents and 7×10^{-6} cm^2/s for glucose. Specific surface for spherical particles is $a = 6/d$ where d is the diameter of the spherical pellet. For mass transfer considerations it is useful to introduce the Sherwood number

$$Sh = \frac{\beta d}{D} \qquad (7.87)$$

and to express the coefficient

$$a\beta = 6 \frac{Sh}{d^2} D \cdot 3{,}600 \quad [\text{h}^{-1}] \tag{7.88}$$

For mildly mixed systems $Sh = 2$ and it increases with Reynolds number which characterizes the degree of mixing. Correlations of Sherwood number for the system geometry and intensity of mixing can be found in the chemical engineering literature. Considering that for the anaerobic microbial system with suspended gel particles intensive mixing is both unnecessary and undesirable, the value of $Sh = 2$ for slow laminar flow around the gel particles may represent a reasonable approximation.

The mathematical model of the growing immobilized-cell process is composed of eight (8) partial differential equations and eight (8) ordinary differential equations characterizing the dynamic behaviour of gel and water phases.

b) *Computer process simulation and discussion*

For the solution of the complex model equation system it is essential to use effective methods for transforming the system of partial differential equations onto Cauchy's problem in ordinary differential equations. Very suitable for this purpose is the orthogonal collocation method derived from approximation of differential operators by orthogonal polynomials. The method, described for instance in the book by Finlayson[6] or Villadsen[45], can be applied in such a way that the first and second partial derivatives by the space coordinate r are replaced by matrix operators A and B:

$$\frac{\partial y}{\partial r} = \frac{2}{d} A y \qquad \frac{\partial^2 y}{\partial r^2} = \frac{4}{d^2} B y \tag{7.89}$$

where y is a vector $(y_1; y_2; \cdots y_n)$ of dependent variables in individual collocation points.

By using this approximation method the system of 8 partial differential equations can be converted into a system of $[8 \times (n+1)]$ ordinary differential equations where n is the number of internal collocation points. This system can subsequently be numerically integrated by i.e. Runge-Kutta method. Following the introduction of matrix operators A and B the boundary conditions for $r = 0$ will become:

$$\frac{2}{d} \sum_{i=0}^{N+1} A_{0,i} y_i = 0 \tag{7.90}$$

for $r = R$:

$$\frac{2D}{d} = \sum_{i=0}^{N+1} A_{N+1,i} \; y_i = \beta \; (y_w - y_{N+1}) \tag{7.91}$$

Through the solution of the above two linear equations the concentrations of sugar, butanol, butyric acid, acetic acid, acetone and ethanol can be determined in the center and on the surface of the gel pellet. The equations can be modified by introducing the Sherwood number as follows:

$$A_{0,0} \; y_0 + A_{0,N+1} \; y_{N+1} = -\sum_{i=1}^{N} A_{0,1} \; y_i \tag{7.92}$$

$$A_{N+1,0} \; y_0 + (A_{N+1,N+1} + \frac{Sh}{2}) \; y_{N+1} = -\sum_{i=1}^{N} A_{N+1} \; y_i + \frac{Sh}{2} \; y_w \tag{7.93}$$

The material balance for internal collocation points is transformed into a system of ordinary differential equations in the following way:

$$\frac{dy_j}{dt} = \frac{4D}{d^2} \left[\sum_{i=0}^{N+1} B_{ji} y_i + \frac{1}{z_j} \sum_{i}^{N+1} A_{ji} y_i \right] + r_g(y) x_j \tag{7.94}$$

for $j = 1, \cdots N$
z_j is a dimensionless coordinate r/R
r_g is the specific rate of product formation or substrate
utilization described by a dependent variable y_j

When solving the model equation, the diffusion coefficient D can be perceived either as a diffusion coefficient of a component in a pure gel phase or as an effective diffusion coefficient in a space filled with microbial cells.

A simplified alternative of the above problem can also be considered whereby the mass transfer rate inside the gel particle is expressed by the maximum diffusion rate in pure gel. A computer program for solution of this problem is presented in Table 7.16 (Appendix). The main program coordinates computations in such a way that the input data are read first, followed by calculations of collocation points and matrix operators A and B including the initial condition of the problem. Subroutines RK4 and ODEC are used for solving the system of ordinary differential equations. Subroutines JCOBI and DFOPR have been transcribed from the Villadsen's book[45] and modified for calculations in single precision. Subroutine RATE and function AMI serve for specification of individual rates of substrate conversion, and subroutine RHS generates the process model with regard to boundary conditions, matrix operators A and B, and the rate of biological conversion for numerical integration using subroutines ODEC and RK4.

The recommended number of N for this case is from 3 to 6, α and β for orthogonal collocation are 1.0 and also as quoted by Villadsen[45].

Results of the simulation of culture growth and product biosynthesis by gel bead-immobilized cells point to a stabilization of production by diffusion. Results of the process simulation with kinetic parameters from Table 7.10 are illustrated by diagrams in Figure 7.35a-e. Figure 7.35a shows the relationship between the concentrations of solvents and biomass respectively, in the flow-through aqueous phase. The outlet sugar concentration initially increases with time as a result of washing out of non-growth medium from the bioreactor. During this period the biomass in the gel phase grows rapidly since there is no inhibition by butanol. The production of butanol is responsible for process stabilization between 15 to 20 h of fermentation when a dynamic equilibrium is established among the growth, product biosynthesis, and substrate utilization parameters. Figure 7.35d shows limitation of the process inside the gel pellet by the substrate. The main resistance can be localized at the level of mass transfer between the gel and water phases. Figure 7.35b shows establishment of the biomass concentration profile during the cell growth in the pellet which appears to be optimal close to the pellet surface where the sugar concentration is maximum while the butanol concentration is the lowest as illustrated by Figure 7.35c. What is valid for butanol is so for the other solvents, acetone and ethanol, as seen in Figure 7.35e. Figures 7.35b-e show the concentration profiles of different compounds in the pellet with time. The simulation results indicate that kinetic equations derived for the conventional continuous-flow culture are not suitable for simulation of the pellet-immobilized, continuous stirred culture which resulted in a realistic biomass concentration profile but solvent concentrations higher than experimentally observed. This disproportion could be explained by a relatively low effect of butanol inhibition $(k_5 = 6)$. The process also appears limited by mass transfer as indicated by relatively flat calculated concentration profiles (Fig. 7.35e).

Since the retention time of gel-immobilized cells in the bioreactor system is much longer compared with the standard flow-through continuous culture, it is reasonable to assume that this results in a significant decrease in the value of inhibition constant K_I (0.5 to 0.833) which may become smaller by an order of magnitude. This effect has been observed with long cell retention times.

The simulation calculations were therefore repeated for the same kinetic parameters, but the batch culture kinetic relationships were applied. Results of this simulation are plotted in Figure 7.36a-e. The concentration profiles for sugar and solvents in the aqueous phase are different from those presented in Figure 7.35 even though it is possible to qualitatively distinguish the periods of wash-out, growth and stabilization of the process. Upon comparison of the two sets of plotted results (Figures 7.35 and 7.36) it can be seen that in the latter case the process of growth and biotransformation is fully controlled at the kinetic level and it is not limited by transfer phenomena.

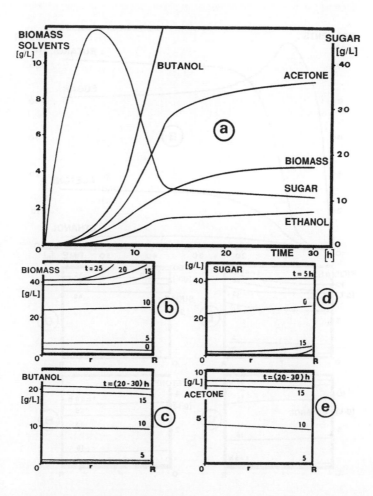

Figure 7.35 (a-e): Immobilized-Cell system in a CSTR with Growth-Supporting Medium (The A-B-E fermentation process):

a) Computer-simulated profiles of the key process variables. This continuous-flow system stabilizes somewhat after hour 15.

b - e) The concentration profiles of selected substances throughout the immobilizing gel beads. The concentration profiles at different fermentation times appear to be constant from the center of the bead (r = 0) to its surface (r =R). Sugar becomes limiting past hour 15, while butanol appears unreasonably high at this time This and other inaccuracies result from a model imperfection when continuous-flow culture kinetic relationships are used.

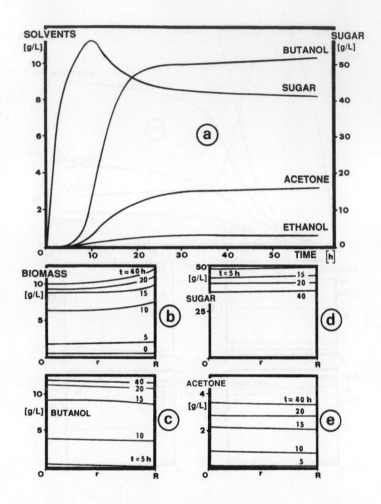

Figure 7.36 (a-e): Immobilized-Cell system in a CSTR with Growth-Supporting Medium in the A-B-E fermentation process. The batch culture kinetic relationships have been used in the model:

    a)   Computer simulated profiles of the key process variables. This continuous-flow system stabilizes after hour 25. The biotransformation appears to be fully controlled at the kinetic level and is not limited by mass transfer phenomena.

    b - e)   The concentration profiles at different fermentation times appear to be relatively constant from the center of the bead (r = 0) to its surface (r = R).

As seen in Figure 7.36, solvent concentrations simulated by this approach correspond well with those reported in experiments by Leung[15] (Figures 7.37 and 7.38). The model can better reflect the behaviour of the actual system when batch model parameters are used. The lack of consistent and applicable experimental data prevents better identification of the model presented here which can thus be considered only as a first approximation. Further fitting of the model to experimental data has not been done at this point for two reasons.

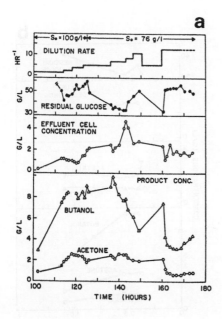
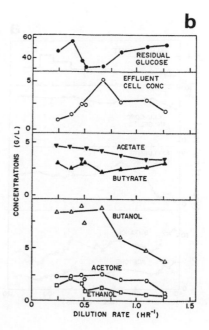

Figure 7.37 (a-b):

Immobilized-Cell system in a CSTR (Continuous Stirred Tank Reactor) with Growth-Supporting Medium. (The A-B-E fermentation process: experimental data after Leung[15]).

a) Time course of the fermentation using immobilized-cell pellets of 3.8 mm diameter.

b) The summary of results for immobilized-cell pellets of 3.8 mm diameter.

Firstly, not enough data were available at the time of model synthesis and, secondly, further work on the model would require a broadfront approach taking into consideration all the basic culture modes such as batch and continuous-flow because, eventually, a model should result which can independently reflect all and any culture modes. Obviously, this approach, as desirable as it may be, would be well beyond the scope of this project. This is also the reason why more extensive computer simulation has not been done with the present model to see just how closely it could reflect the available experimental data. This exercise, requiring substantial computer time, could be carried out as a further extension of this work by an interested student of the process.

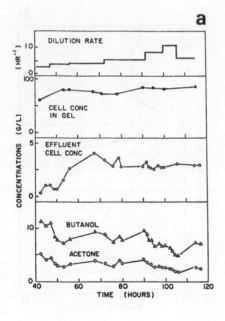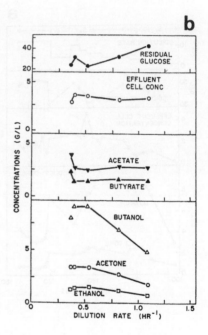

Figure 7.38 (a-b):

Immobilized-Cell system in a CSTR (Continuous Stirred Tank Reactor) with Growth-Supporting Medium. (The A-B-E fermentation process: experimental data after Leung[15]).

a) Time course of the fermentation using immobilized-cell pellets of 5.0 mm diameter.

b) The summary of results for immobilized-cell pellets of 5.0 mm diameter.

While the current model is capable of reflecting the culture trends apparent from Leung's[15] experiments, it appears that for simulation of the gel pellet-immobilized stirred continuous-flow culture of *C. acetobutylicum* it is better to use kinetic data derived from batch culture experiments because they better reflect the history and age structure of the microbial population with regard to establishing a dynamic equilibrium between the environment and its phases including physiological changes resulting in changes of inhibition constant values.

The second significant lesson derived from these simulation studies is based on pinpointing the process bottleneck which is associated with either the interface mass transfer or with the biosynthesis kinetic limitations. To consider the slow-down effect of the intra-pellet diffusion is probably not important because, as indicated by simulation studies for high biomass concentrations (80-100 g/L gel) in certain locations within the pellet, compared to the overall process kinetic rate, the gel diffusional limitations have a very small impact. The latter effect should be reckoned with, however, when the process simulation is carried out for processes where a continuous microbial film is grown. The acetone-butanol fermentation process, however, does not belong to this category. Its productivity benefits greatly from preservation and immobilization of the biomass "catalyst" in the gel phase where it can be maintained at very high concentrations.

Certain amounts of cells invariably "leak" from the gel phase and may slightly complicate the simulation-experiment comparison for the system, however, it is not considered a significant drawback. A high concentration of cells, attainable in the immobilized cell system, enables much better utilization of the bioreactor volume and the entire system, which by its nature corresponds best with the natural conditions under which *C. acetobutylicum* prospers.

Since an industrial application of the stirred bioreactor with gel pellet-immobilized cells would be a rather atypical one, the following two Sections will focus on cases of bioreactors with a fixed bed of particle-immobilized cells. Two media alternatives will be analyzed, namely a growth-supporting one and a growth-nonsupporting production medium. Considering that the effect of intraparticle diffusion of substrate and product does not have a significant impact on the overall process performance as is the case for process kinetics and interface mass transfer, the model can be simplified by assuming an ideally mixed gel phase where the mixing is due to intensive diffusion within the pellet.

7.4.2 IMMOBILIZED-CELL SYSTEM
WITH NON-GROWTH MEDIUM

a) *Model development*

By immobilization of microbial cells in a gel matrix a new phase is introduced into the fermentation process bioreactor represented by the gel particles. When modeling such a system, the rate of interphase mass transfer has to be considered in addition to the reaction rates. In experiments carried out with growth non-supporting medium the microbial cells are basically used as a catalyst and the modeling problems can be approached from the basis well established by reactor engineering studies. For the immobilized-cell reactor design the relevant reaction kinetics has to be evaluated first. For this purpose experimental data on immobilized cell A-B-E process will be considred as presented by Haggström[8,9,10]. A mathematical model formulated for a two phase reactor[15] has to contain the interphase mass transfer rate as an additional parameter. Estimation of the water-gel mass transfer time constant (τ_G) for all components can be based on experimental data by Leung[15] giving the range of $\tau = (250\text{-}300)$sec. This constant will be included in the mathematical process model assuming that the catabolic two-phase process kinetics is identical with that for a single-phase system. The material balances for a gel phase of volume V_G can be written as follows:

Biomass material balance in the gel phase:

$$\frac{\mathrm{d}X_G}{\mathrm{d}t} = -k_2\, X_G\, B_G \tag{7.95}$$

Sugar material balance in the gel phase:

$$\frac{\mathrm{d}S_G}{\mathrm{d}t} = \frac{1}{\tau_G}(S_W - S_G) - k_3 S_G X_G - k_4 \frac{S_G X_G}{S_G + K_s} \tag{7.96}$$

Butyric acid material balance in the gel phase:

$$\frac{\mathrm{d}(BA)_G}{\mathrm{d}t} = \frac{1}{\tau_G}[(BA)_W - (BA)_G] + k_5 \frac{S_G K_I}{K_I + B_G} X_G -$$

$$- k_5 \frac{(BA)_G}{(BA)_G + K_{BA}} X_G \tag{7.97}$$

Acetic acid material balance in the gel phase:

$$\frac{d(AA)_G}{dt} = \frac{1}{\tau_G}[(AA)_W - (AA)_G] + k_8\frac{S_G}{S_G + K_S}\frac{K_I}{B_G + K_I}X_G -$$

$$- k_9\frac{AA_G}{AA_G + K_{AA}}X_G \qquad (7.98)$$

Butanol material balance in the gel phase:

$$\frac{dB_G}{dt} = \frac{1}{\tau_g}(B_W - B_G) + k_7 S_G X_G - 0.831\, k_5\frac{S_G K_I}{K_I + B_G}X_G +$$

$$+ 0.831\, k_6\frac{(BA)_G}{(BA)_G + K_{BA}}X_G \qquad (7.99)$$

Acetone material balance in the gel phase:

$$\frac{dA_G}{dt} = \frac{1}{\tau_G}(A_W - A_G) + k_{10}\frac{S_G}{S_G + K_S}X_G - 0.5\, k_8\frac{S_G}{S_G + K_S}\frac{K_I}{K_I + B_G}X_G +$$

$$+ 0.5\, k_9\frac{AA_G}{AA_G + K_{AA}}X_G \qquad (7.100)$$

Ethanol material balance in the gel phase:

$$\frac{dE_G}{dt} = \frac{1}{\tau_G}(E_W - E_G) = k_{11}\frac{S_G}{S_G + K_S}X_G \qquad (7.101)$$

The following material balances pertain to the water phase of volume V_W:

$$V_W\frac{dS_W}{dt} = -\frac{V_G}{\tau_g}(S_W - S_G) \qquad (7.102)$$

Butyric acid material balance in the water phase:

$$V_W\frac{d(BA)_W}{dt} = -\frac{V_G}{\tau_G}[(BA)_W - (BA)_G] \qquad (7.103)$$

Acetic acid material balance in the water phase:

$$V_W\frac{d(AA)_W}{dt} = -\frac{V_G}{\tau - G}[(AA)_W - (AA)_G] \qquad (7.104)$$

Butanol material balance in the water phase:

$$V_W \frac{dB_W}{dt} = -\frac{V_G}{\tau_G} (B_W - B_G)$$ (7.105)

Acetone material balance in the water phase:

$$V_W \frac{dA_W}{dt} = -\frac{V_G}{\tau_G} (A_W - A_G)$$ (7.106)

Ethanol material balance in the water phase:

$$V_W \frac{dE_W}{dt} = -\frac{V_G}{\tau_G} (E_W - E_G)$$ (7.107)

Experimental data on the gaseous phase of the A-B-E system were not reported in the relevant literature[8,9,10,15]. However, for completion of the mathematical model presented here, the gas evolution rates could be estimated from the batch process kinetic constants established in Section 7.1. The mass rates of gas evolution could be then defined as:

$$G'_{CO_2} = k_{12} \frac{S_G}{S_G + K_S} V_G X_G$$ (7.108)

$$G'_{H_2} = k_{13} \frac{S_G}{S_G + K_S} V_G X_G$$ (7.109)

The derivation of the model presented above was based on a two-phase heterogeneous stirred reactor while complete ideal mixing was assumed in both phases. In fact, there could easily be concentration gradients occurring in the gel phase at higher reaction rates. However, effects of these concentration gradients could be implicitly included as an "effectiveness factor" in individual kinetic constants. Diffusion rates inside the gel pellets during the non-growth regime are almost an order of magnitude higher than the reaction rates. This fact enables good approximation of the dynamic behaviour of the gel phase by assuming a non-gradient concentration transfer regime. Results of the identification calculations indicate that the kinetic constants range within the values determined earlier for the parameters of the A-B-E process kinetics (compare with Tables 7.7 and 7.11). The only difference is in the value of constant k_2 characterizing the cell decay rate. In the non-growth medium the capacity of cells to renew their life function is apparently decreasing which is also given by the fact that removal of nitrogen source from the medium shifts its buffering capacity into the acidic region accelerating thus the process of cell lysis.

b) *Computer process simulations and discussion*

Simulation calculations were carried out for the culture conditions reported by Haggström[8,9,10] and kinetic constants of the process simulation model were quantitatively expressed. When considering a larger volume of incompatible experimental data, values of some parameters were expressed as ranges for constants evaluated for different initial conditions of the cultivation (Table 7.17). The simulation and model identification procedure used slightly modified software described in previous paragraphs.

During the biosynthesis of neutral solvents in the immobilized cell reactor using non-growth medium, the immobilized cell activity may be lost at least in part due to the toxic effect of the products.

This kind of a process does not allow for the achievement of higher concentrations of solvents and compared to the standard batch process does not bring any special advantages. The continuous-flow based immobilized cell process[9] did not exhibit a long-term steady state that was interesting from the solvent production point of view. The immobilized-cell process is similar to the chemical catalytic processes with catalyst deactivation. Figure 7.39 shows simulation results for a repeated batch solvent production process where fresh medium with 20 g glucose/L replaces the used one every 24 hours. For better clarity and comparison with Haggström's[8] results (Figure 7.40) Figure 7.39 presents only butanol concentration in the water phase and biomass concentration in the gel phase. In the initial sequential batch fermentations, Haggström[8] reached higher butanol productivity which, however, was followed by a rapid decrease of biological activity in the gel phase where the biomass became significantly deactivated. The biomass decay resulted in a decreased solvent production which in turn led to a lowered solvent concentration resulting in a small rate of cell lysis and thus relative stabilization of the process. The actual butanol and biomass concentrations are shown in the Figure because in the Haggström's original work[10] the low butanol production is plotted as a relative ratio of the first batch run to the rest. The process simulation model explains on a process kinetics level the phenomenon designated in the cited work as productivity stabilization.

From the systems analysis point of view, the immobilized cell A-B-E process alternative using non-growth medium is the least advantageous technological variation of the process. The complications encountered with the preparation of the immobilized-cell reactor which requires special techniques and anaerobic handling should be well counterbalanced by higher solvent productivities of solvents and higher conversion rates of high-concentration

Figure 7.39:

Immobilized-Cell system in a CSTR with
Non-Growth Medium

(The A-B-E fermentation process):
Simulation results for a repeated batch
solvent production where fresh medium
with 20 g/L replaces the used one every
24 hours. Only profiles for the butanol
concentration in the liquid phase (B_W)
and the biomass concentration in the gell
phase (X_G) are shown.

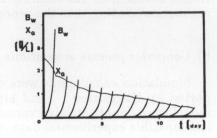

Figure 7.40:

Immobilized-Cell repeated batch culture
in a mixed bioreactor with Non-Growth
Medium containing glucose, butyric acid
and "inorganic salts".
(The A-B-E fermentation process
experimental data after Haggstrom[8]).

a) The decreasing butanol production
   rate over a period of time; the
   medium was being replaced "in
   intervals" during the 11-day
   culture period.

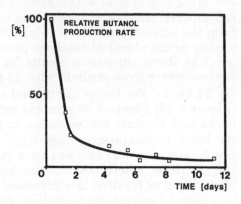

b) The butanol concentration profile
   over three non-growth medium
   replenishment steps. The biomass
   concentration in alginate beads
   (X_G) was not assessed. The
   original   spore   inoculum
   immobilized in the beads was first
   incubated for 24 hours in a
   complete growth medium.

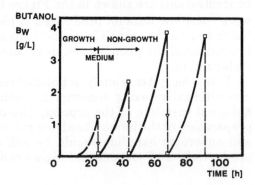

sugar feeds. Elimination of the possibility of regenerating the natural bio-catalyst by maintaining non-growth culture conditions leads to extreme cell deactivation. Considering the well known fact that higher solvent productivity is limited by the resistance of the microorganism to growth-inhibiting metabolic end-products, the use of non-growth medium goes entirely against the requirements for the optimal process performance.

Leung in his study[15] of the immobilized *Clostridium acetobutylicum* culture showed understanding of the importance of culture self-reactivation when a growth medium was used. This resulted consistently in a high utilization of the reaction volume and of the reaction substrate which was reflected in a high hourly rate of the neutral solvent production. Analysis of experimental results presented by Leung[15] was presented in the preceding Section 7.4.1.

TABLE 7.17

A-B-E Process Kinetic Model Parameters for the Immobilized-Cell
Non-Growth (Production) Medium System

Kinetic constants:

$$k_1 = 0 \text{ (non-growth)}$$
$$k_2 = 0.008$$
$$k_3 = 0.024$$
$$k_4 = 0.6$$
$$k_5 = 0.0135\text{-}0.028$$
$$k_6 = 0.08\text{-}0.18$$
$$k_7 = 0.02\text{-}0.035$$
$$k_8 = 0.18$$
$$k_9 = 0.01$$
$$k_{10} = 0.08$$
$$k_{11} = 0.026^*$$
$$k_{12} = 0.6^*$$

Other constants:

$$K_I = 0.83$$
$$K_S = 2$$
$$\tau_G = 0.07\text{-}0.083$$
$$K_{AA} = 0.5$$
$$K_{BA} = 0.5$$

* Taken from the batch culture kinetic model.

7.4.3 IMMOBILIZED-CELL SYSTEM ON NON-GROWTH MEDIUM IN A TUBULAR FIXED-BED REACTOR WITH AXIAL DISPERSION

a) *Model development*

When formulating the process model for the immobilized cell system the effect of diffusion inside the gel particle with biomass concentrations lower than 15 g/L of gel was demonstrated in the previous Section. The dynamic behaviour of the immobilized cell system has to be divided into 2 constituent phases as shown in Sections 7.4.1 and 7.4.2.

Formulation of the system model will be similar to the approach taken for a two-phase adiabatic reactor reported elsewhere[11]. The two constituent phases, gel and water, will be considered respectively. The following equations are the material balances for process species in the gel phase at an arbitrary coordinate ℓ :

Biomass:

$$\frac{\partial X_G}{\partial t} = -k_2 X_G B_G \tag{7.110}$$

Sugar:

$$\frac{\partial S_G}{\partial t} = -\left[k_3 S_G + \frac{k_4}{S_G + K_s}\right] X_G + \frac{1}{\tau_G}(S_W - S_G) \tag{7.111}$$

Butyric Acid:

$$\frac{\partial (BA)_G}{\partial t} = k_5 \frac{S_G K_I}{K_I + B_G} X_G - k_6 \frac{(BA)_G}{(BA)_G + K_{BA}} X_G +$$

$$+ \frac{1}{\tau_G}[(BA)_W - (BA)_G] \tag{7.112}$$

Acetic Acid

$$\frac{\partial (AA)_G}{\partial t} = k_8 \frac{S_G}{S_G + K_s} \frac{K_I}{K_I + B_G} X_G - k_9 \frac{(AA)_G}{(AA)_G + K_{AA}} X_G +$$

$$+ \frac{1}{\tau_G}[(AA)_W + (AA)_G] \tag{7.113}$$

Butanol:

$$\frac{\partial B_G}{\partial t} = \left[k_7 S_G - 0.831 k_5 S_G \frac{K_I}{K_I + B_G} + 0.831 k_6 \frac{(BA)_G}{(BA)_G + K_{BA}} \right] X_G +$$

$$+ \frac{1}{\tau_G}(B_W - B_G) \tag{7.114}$$

Acetone:

$$\frac{\partial A_G}{\partial t} = \left[k_{10} - 0.5 k_8 \frac{K_I}{K_I + B_G} \right] \frac{S_G X_G}{S_G + K_s} + 0.5 k_9 \frac{(AA)_G}{(AA)_G + K_{AA}} X_G +$$

$$+ \frac{1}{\tau_G}(A_W - A_G) \tag{7.115}$$

Ethanol:

$$\frac{\partial E_G}{\partial t} = k_{11} \frac{S_G}{S_G + K_s} X_G + \frac{1}{\tau_G}(E_W - E_G) \tag{7.116}$$

Material balance of the water phase including the effect of backmixing characterized by a diffusion effective coefficient D_E will be the same for all the compounds (sugar, butanol, acetone, ethanol, butyric acid, acetic acid):

$$\frac{\partial C_W}{\partial t} = D_E \frac{\partial^2 C_W}{2\ell^2} - u \frac{\partial C_W}{\partial \ell} - \frac{(1 - \epsilon)}{\epsilon \tau_G}(C_W - C_G) \tag{7.117}$$

where ϵ is the free volume of the gel particle bed.

For the whole system the following initial and boundary conditions will be used:

$$t = 0; \quad X_G = X_0$$
$$S_W = B_W = A_W = E_W = BA_W = AA_W = 0$$
$$S_G = B_G = A_G = E_G = BA_G = AA_G = 0$$

$t \geq 0$;   for the end of the tubular reactor $(\ell = L)$:

$$\frac{\partial C_W}{\partial \ell} = 0 \tag{7.118}$$

for the beginning of the tubular reactor $(\ell = 0)$:

$$-D_E \frac{\partial C_W}{\partial \ell} = u(C_W^o - C_W) \tag{7.119}$$

These boundary conditions of the Danckwerts type are most frequently used for the reactor design. Results of the model identification calculations indicate that the kinetic constant value ranges are similar to those determined for the batch version of the process analyzed previously. The only difference is in the value of constant k_2 characterizing the cell decay rate which is more pronounced in the non-growth medium.

The mathematical model of the system is represented by a set of partial differential equations.

b) *Computer process simulations and discussion*

For the solution of the set of model equations, it is possible to use a computer subroutine DSS-2 available in the McGill University (Montreal) computer program library, or a method of orthogonal collocation for this case. According to the number of collocation points, the set of 13 partial differential equations can be transferred into a set of $(N \times 13-12)$ ordinary differential equations. The boundary conditions can be transcribed by using matrix operators of the orthogonal collocation method for one dimensional cartesian ordinates into the following form:

$$\sum_{i=0}^{N+1} A_{N+1,i} \; C_{Wi} \; = \; 0 \tag{7.120}$$

$$-\frac{D_E}{L} \sum_{i=0}^{N+1} A_{0,i} \; C_{Wi} \; = \; u \, (C^\circ - C_{W0}) \tag{7.121}$$

Following the numerical expression of values for C_{W0} and $C_{W(N+1)}$ and introduction of the Peclet number for turbulent diffusion in a layer of particles of diameter d_p and the layer length L, the following relationships can be obtained:

$$A_{NM,0} \; C_{W0} + A_{N+1,N+1} \; C_{W,N+1} \; = \; \sum_{i=0}^{N} A_{N+1,i} \; C_{Wi} \tag{7.122}$$

$$\left(\frac{PeL}{d_p} - A_{0,0}\right) C_{W0} - A_{0,N+1} \; C_{W,N+1} \; = \; +\sum_{i=1}^{N} A_{0,i} \; C_{Wi} + \frac{PeL}{d_p} \, C^\circ \tag{7.123}$$

Equations for the water phase can be modified in a similar manner:

$$\frac{dC_{Wj}}{dt} = \frac{D_G}{L^2} \sum_{i=0}^{N+1} B_{ji}\, C_{Wi} - \frac{u}{L} \sum_{i=0}^{N+1} A_{ji}\, C_{Wi} - \frac{1-\epsilon}{\epsilon}\frac{1}{\tau_G}(C_{Wj}-C_{Gj}) \quad (7.124)$$

Upon introducing the Peclet number and dilution rate $D = \frac{u}{L}$ the above equation is transferred into a form more suitable for solving it:

$$\frac{dC_{Wj}}{dt} = D_G \sum_{i=0}^{N+1}\left[\frac{1}{Pe}\frac{d_p}{L}B_{ji} - A_{ji}\right] C_{Wi} - \frac{1-\epsilon}{\epsilon}\frac{1}{\tau_G}(C_{Wj}-C_{Gj}) \quad (7.125)$$

Equations describing the gel phase phenomena are formally transformed into a set of ordinary differential equations at individual collocation points. This resulting set of equations can be numerically solved by a Runge-Kutta method for initial conditions specified earlier.

The computer program for solving this mathematical model of the process is presented in Table 7.18 (Appendix). The structure of this program, based on the method of orthogonal collocation, is similar to that used in Section 7.4.1 with only a change introduced in the model formulation in subprogram RMS. According to the number of given collocation points, 13 N+14 ordinary differential equations is to be solved simultaneously by using subprograms ODEC and RK4. Programs JCOBI and DFOPR are used for calculation of matrix operators A and B. In subprogram RATE the rates of substrate utilization and of acid production as well as of solvents synthesis are expressed. The Peclet number, characterizing the backmixing, can be quantified e.g. by Edwards-Richardson correlation[5]:

$$\frac{1}{Pe} = \frac{0.73\epsilon}{Re\cdot Sc} + \frac{0.5}{1+\frac{9.7\epsilon}{Re\cdot Sc}} \quad (7.126)$$

For the *C. acetobutylicum* cultivation the Peclet number value is (2-2.3), meaning that backmixing will be in shorter layers with $L/d_p \geq 50$.[11] For longer layers the bioreactor can be modelled as the one with a piston flow. Figure 7.41 shows an example of a simulated bioreactor performance for immobilized growth kinetics discussed previously, but under non-growth conditions ($\mu = 0$). It can be seen that during the first day of operation pure medium is being washed out of the system. After 20 hrs almost maximum conversion of sugar into solvents is observed, whereby butyric acid is converted into butanol. Following the first day of operation the solvent production in the bioreactor is stabilized. Butanol toxicity results in decreasing biomass at the end of the bioreactor which in turn results in increasing sugar and decreasing solvent concentrations. Changing the bioreactor geometry from $L/d = 20$ to $L/d = 40$, keeping the retention time constant, results in practically no variations in solvent productivity and sugar consumption.

In terms of technological process parameters, an important role is played by the overall bioreactor retention time and by the interface mass transfer rate. For higher values of L/d (≥ 20), the bioreactor behaves as one with piston flow in the flow zone of the layer. Figure 7.41c-g presents the calculated profiles for the biomass, sugar, and products concentrations in the water phase. Figure 7.41b indicates the cell lysis in the second half of the bioreactor which results from butanol toxicity as sugar is being gradually converted into this solvent product (Figures 7.41c-d), the highest concentration of butanol being in the "downstream" section of the bioreactor length. This is also similar for ethanol (Figure 7.41f). Figure 7.41g shows the butyric acid concentration profile along the bioreactor. Since butyric acid is an intermediate compound in butanol biosynthesis, its concentration increases first before it decreases due to its conversion and due to the effect of butanol toxicity on microbial cells which cease producing it.

The application of an immobilized-cell fixed-bed reactor system for an anaerobic fermentation process makes it possible to spread the production biosynthetic relationships into space and eventually to control the process by local siphoning-off of the broth with products and by eventual adding of the substrate. Another possible control strategy is in changing the dilution rate as a function of sugar concentration. Since the bioreactor behaves more like the one with piston flow, decreasing the dilution rate implies increasing the degree of conversion. Figure 7.42 presents the simulation case for a simple control strategy where the dilution rate was decreased by a factor of 0.9 when the glucose concentration in the exiting stream exceeds 2 g/L. In this simulation case the condition for decreasing the dilution rate was tested every hour. The results are showing that this interval is too short because at hour 82, 108, and 146 the system was over-regulated by an excessive decrease of the dilution rate. This shortcoming could be overcome by either increasing the dilution rate controlling factor from 0.9 to 0.95 or more, or by increasing the sample withdrawal interval from 1 hour to 2 hours or more. This measure will result in suppressed losses of sugar substrate in exchange for slightly decreased productivity of the system. There will be practically no change in the product stream composition, as seen in Figure 7.42.

The simulation results show that the immobilized-cell bioreactor employing non-growth medium could be maintained in productive operation with realistic biomass concentration for approximately a week when a simple control strategy is applied.

An interesting case would be using a bioreactor with growth-supporting medium which is analyzed in the next Section.

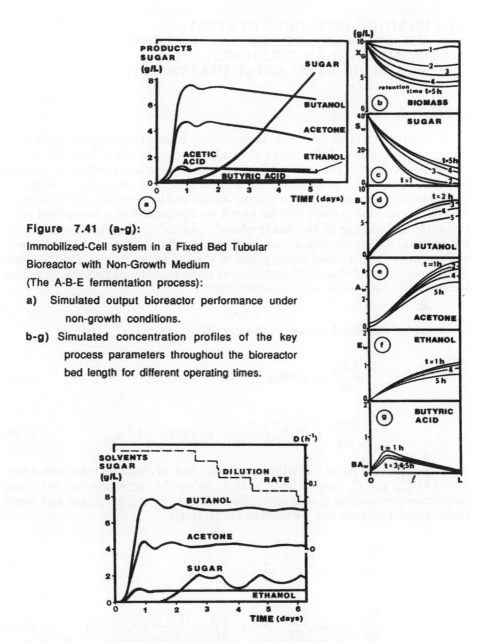

Figure 7.41 (a-g):

Immobilized-Cell system in a Fixed Bed Tubular
Bioreactor with Non-Growth Medium
(The A-B-E fermentation process):
a) Simulated output bioreactor performance under
    non-growth conditions.
b-g) Simulated concentration profiles of the key
    process parameters throughout the bioreactor
    bed length for different operating times.

Figure 7.42 : Immobilized-Cell system in a Fixed Bed Tubular Bioreactor with
                Non-Growth Medium (The A-B-E fermentation process).
Simulated output bioreactor performance with a simple step-control (decrease) of the
dilution rate whenever the output sugar concentration exceeds 2 g/L.

7.4.4 IMMOBILIZED-CELL SYSTEM ON GROWTH-SUPPORTING MEDIUM IN A TUBULAR FIXED-BED REACTOR WITH AXIAL DISPERSION

a) *Model development*

A fixed-bed bioreactor with immobilized cells producing solvents on a non-growth medium behaves as a catalytic reactor with de-activation of the catalyst by the product. When growth-supporting medium is applied in the system an atypical behaviour can be expected since the growing microorganisms can to a certain degree adapt to the concentration changes. Simulation of this system will be based on the model of a two-phase reactor with backmixing in the water phase[11] and on the batch kinetics model for growth, sugar bioconversion and product biosynthesis derived previously. Since the detailed description of the model equations for the basic non-growth immobilized-cell system was done earlier in this text, the mass balances in this section will be abbreviated.

Material balance for biomass in the gel phase:

$$\frac{\partial X_G}{\partial t} = 0.56\,(y_G - 1) - k_2 X_G B_G \qquad (7.127)$$

$$\frac{\partial y_G}{\partial t} = \left[k_1 S_G \frac{K_I}{K_I + B_G} - 0.56(y_G - 1) \right] y_G \qquad (7.128)$$

If concentrations of the other products and of the substrate are designated as C_W and C_G respectively, with r_G being the corresponding gel-phase respective production and consumption rates, then the gel-phase and water phase mass balances can be written respectively:

$$\frac{\partial C_G}{\partial t} = r_G X_G + \frac{1}{\tau_G}\,(C_W - C_G) \qquad (7.129)$$

$$\frac{\partial C_W}{\partial t} = D_E \frac{\partial^2 C_W}{\partial \ell^2} - u \frac{\partial C_W}{\partial \ell} - \frac{(1 - \epsilon)}{\epsilon \tau_G}\,(C_W - C_G) \qquad (7.130)$$

with initial conditions:

$$t = 0\,; \quad C_W = C_G = 0\,; \quad X_G = X_o\,; \quad y_G = 1.0$$

and boundary conditions:

$$\ell = 0 \qquad -D_E \frac{\partial C_W}{\partial \ell} = u\,(C_W^\circ - C_W) \qquad (7.131)$$

$$\ell = L \qquad \frac{\partial C_W}{\partial \ell} = 0 \qquad (7.132)$$

These relationships apply for all substrate and product concentrations in the gel and water phases. The process mathematical model is represented by a set of 14 partial differential equations. For the process computer simulation a variant of the orthogonal collocation method was used.

The corresponding computer program is presented in Table 7.16. The sub-program RHS was supplemented by differential equations from Section 1 which describe the growth kinetic relationships and by real function AMI which is identical with the function used for simulation of the stirred bioreactor with growing immobilized cells in Section 7.4.1. The main program was expanded by initial conditions characterizing the growth dynamics. According to what is the chosen number of collocation points N, the system of $14N + 16$ ordinary differential equations were solved by a Runge-Kutta method of the 4th order.

Figure 7.43 presents the results of a computer simulation of the bioreactor with immobilized cells on growth-supporting medium for the same technological process parameters as applied in the simulation with non-growth medium presented in Figure 7.41. Upon comparison of the two cases it can be seen that during the initial period of the process (fermentation day 1) the process relationships are influenced by the washout of medium from the bioreactor. Following the initial period, the character of relationships changes as a result of the *C. acetobutylicum* cell growth. A dynamic equilibrium is slowly established in the bioreactor which is characterized by a high degree of sugar conversion to solvents as seen from Figure 7.43 presenting time-concentration profiles for sugar and solvents. After 80 hours a continuous-flow operation performance of the bioreactor is stabilized.

Experimental data for this system were obtained from experiments conducted at the Canadian National Research Council in Ottawa[44]. When the simulation results are compared with experimental data[44], the prediction of the process behaviour based on the batch-culture data exhibits a difference in the acetone and butanol concentrations in that the exit stream of the bioreactor had approximately 1 g butanol/L more and 1 g acetone/L less. Concentrations of ethanol and acetic acid, including the degree of sugar conversion, agree with the reported experimental data[44]. Considering the variability of *C. acetobutylicum* and that the pH in the immobilized-cell reactor was not controlled, agreement of the simulation results with those of the experiment is very good since the overall mass yield of solvents and acetic acid was exactly determined, including the sugar conversion at the exit port of the bioreactor.

Figure 7.43b-g presents simulated concentration profiles for biomass, fermentation products, and for the substrate along the bioreactor axis. As seen from Figure 7.43, the biomass gradually grows in the direction of the substrate gradients. A similar effect was observed for the growth simulation in the gel pellet under the diffusion limitation as discussed in the previous section. In its result, this phenomenon could even lead to the growth clogging of the reactor inlet region at lower dilution rates. The substrate concentration profile along the bioreactor axis is shown in Figure 7.43c. In the first fermentation phase ($t = 10$ hrs), the profile is almost linear because the process is not limited by substrate. During a prolonged bioreactor functioning the substrate concentration profiles were simulated in Figure 7.43c, showing a decreasing trend in time. The solvent concentration profiles (Figure 7.43d-f) are all very alike, being characterized by a steep increase in the first part of the bioreactor whereby they asymptotically approach the exit concentration values. The simulated butyric acid concentration profile depicted in Figure 7.43g indicates a rapid biosynthesis of the acid in the starting process phase followed by its gradual suppression by butanol produced.

In general, it may be concluded that the computer simulation model presented here for the A-B-E process with a fixed-bed immobilized-cell bioreactor is suitable for adequately representing the behaviour of the process. The model enables prediction of the bioreactor performance at different configurations and dilution rates. For values of $L/d \geq 20$ the degree of conversion for the same dilution rate, but different geometries, does not differ. The key process parameter is the dilution rate which can be used for controlling the process. The bioreactor with a growing culture immobilized in a gel phase behaves differently from chemical reactors since the dynamics of culture adaptation along the distance coordinate plays an important role.

This adaptation could be affected mainly by the dilution rate in the system. Even though this rate is not equal to the specific growth rate (μ), as is usually the case with conventional stirred continuous-flow culture systems, it has a significance of a convective flow between the system and its surroundings.

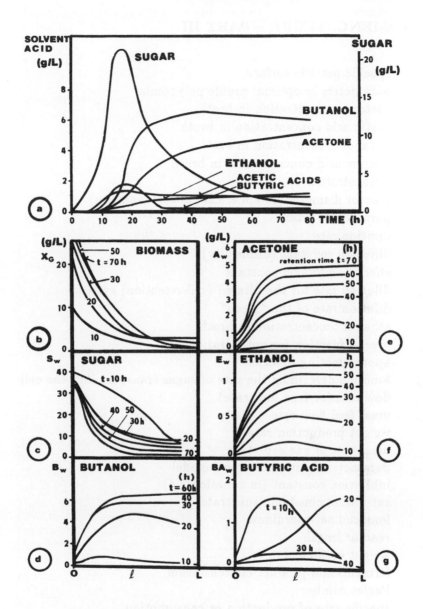

Figure 7.43 (a-g): Immobilized-Cell system in a Fixed Bed Tubular Bioreactor with Growth-Supporting Medium (The A-B-E fermentation process):

a) Simulated output performance of a bioreactor fed with growth supporting medium.

b-g) Simulated concentration profiles of the key process parameters throughout the bioreactor bed length for different operating times.

7.5 NOMENCLATURE – PART III

a	specific particle surface
A_i	parameters in optimal profile polynomial
A	acetone concentration in broth
AA	acetic acid concentration in broth
B	butanol concentration in broth
BA	butyric acid concentration in broth
C	concentration(s)
d	reactor diameter (or spherical particle diameter)
d_p	particle diameter
D	dilution rate
D	diffusion constant (diffusivity)
D_E	effective diffusion constant
D_{FILT}	dilution rate for the filtered (cell-retention) system
D_2	dilution rate of the "bleed"
E	ethanol concentration in broth
f_{OPT}	overall function for optimization
f_i	approximating function
$f(S)$	function describing the rate of sugar transport into the cell
F	flow rate (or feed flow rate)
F'	mass feed flow rate
G	weight production rate
G'	mass rate of gas evolution
k_i, k_i'	parameters in mathematical model
K_I	inhibition constant (in equation 7.4)
K_S	saturation constant (substrate)
ℓ	longitudinal coordinate
L	reactor length
P	weight coefficient
P	fermentation product concentration
Pe	Peclet number
q	specific rate of production or consumption
$q(\xi)$	age category
$Q(t)$	mean metabolic activity value for cells in the population characterized by $x(\xi)$, $q(\xi)$
r	radial distance
r_s	substrate utilization rate

r_p	product synthesis rate
R	radius of immobilization bead
Re	Reynolds number
S	substrate concentration in the broth
Sc	Schmidt number
Sh	Sherwood number
S_0	substrate concentration in the feed stream or initial sugar concentration in a batch
SSR	sum of squares of residues (Eq. 7.20)
SSW	sum of squares of weighted residues (Eq. 7.21)
t	time
t_i	time series
u	longitudinal flow velocity
V	volume
V_0	starting batch volume
V_f	final broth volume
W_j	weighing coefficient
X	biomass concentration
y	yield coefficient
$(y$	$RNA/RNA_{min})$
y_i	measured value
y_j	dependent variable for the specific rate of product formulation or substrate utilization
z_j	dimensionless coordinate r/R

8. CONCLUSION

The "Case Study" has been developed here as a comprehensive example of mathematical modeling of different modes of a bio-process. The operation modes of a bioreactor considered spanned a whole spectrum from a simple and most conventional batch culture process to the most contemporary process using immobilized cells technology. The acetone-butanol fermentation process selected for this exercise is not a widely utilized one by the industry any more. However, for the study purposes, it represents a good example of a fairly complex fermentation process where the substrate is converted into a variety of products, namely the solvents: acetone, butanol and ethanol (minute amounts of other end-metabolites which are also known to have been detected in this fermentation such as e. g. acetoin, lactic acid or a by-product polysaccharide have not been considered in the modeling attempts) and two fermentation off-gasses: hydrogen and carbon dioxide. The complex metabolic conversion process features also at least two major intermediates, acetic and butyric acids, which accumulate and become partially reabsorbed during the latter stages of the fermentation. The spore-forming anaerobic bacterium *Clostridium acetobutylicum* used in the fermentation process is known for its intriguing physiology and metabolic behavior.

All considered, the selected "case study" fermentation process represents a challenging bio-conversion system which is relatively difficult to mathematically model. Only recently has the combination of accumulated knowledge of the relevant biochemical pathways, the advances in the mathematical modeling methodology and the increasing availability of extremely powerful computer hardware and sophisticated software allowed meaningful attempts in mathematical modeling of a complex fermentation process such as the one chosen here. The "case study" developed here in some detail can represent a good basic pattern of how a fermentation process can be modelled. While there are many other bio-processes of industrial or scientific interest, the modeling methodology outlined here can be readily applied in describing these, in many cases, even much simpler processes. In general terms, a fermentation proces featuring intermediate compounds and liquid as well as gaseous products of the substrate conversion represents what can be considered as a typical fermentation process. In some cases the intermediates are not found or the final metabolites are not as many. These cases would only become simplified versions of the fermentation process chosen and analyzed here in the "case study". The methodology of mathematical modeling for different fermentation processes will be, for all practical purposes, identical and even simpler than the example selected here. This by no means implies that it always has to be the case. Naturally, a good biochemical knowledge of the process and its major pathways is an important prerequisite.

When the culture physiology is of a pronounced importance, the mathematical models describing such a system may become rather complex being very highly structured in order to be appropriately sensitive to those "internal" culture features considered important. That invariably leads to opening a whole new "pandora box" of what has been generally termed the "physiological state" of the culture, the main problem being in appropriately reflecting it in the model structure by using well chosen "markers" of the physiological state of the culture. The search for these "markers", or measurable relevant key parameters, has been going on for some time. The "case study" in this volume is showing one of the first ones used successfully, namely the cellular RNA concentration. However, this particular parameter appears to be mainly related to the growth activities of the culture. When it comes to describing the culture physiology as it relates to and is reflected in the metabolic product synthesis, particularly during the growth non-associated" product formation phase, the suitable "markers" available are relatively few and far between.

The most recent advances in culture fluorometry have opened new possibilities for *in situ* determination of mainly the culture NADH fluorescence. The culture NADH fluorescence probes have been used both in industry as well as in research laboratories to monitor aspects of the culture behavior. While the NADH fluorescence signal can be relatively reliably obtained from the bioreactor, the interpretation of the signal is not always that easy and straight forward because its level mainly depends on both the culture "quantity" in terms of the biomass concentration as well as on the culture "quality" in terms of its metabolic activity and associated leves of NADH in the cell. This dilemma is well illustrated in the still somewhat inconclusive recent research efforts relevant to the Case Study in this book. The interpretations of the NADH fluorescence signal from the *C. acetobutylicum* cultures[25, 26, 27, 32, 33] have been disputed in the recent work of Srivastava and Volesky[34, 35, 36]. The latter work was specifically conducted for the purpose of incorporating the NADH fluorescence in the mathematical model of the fermentation as a "physiological state marker". The results of these attempts are presented in a series of the most recent research papers by Srivastava and Volesky published in 1991 [37, 38, 39, 40, 41, 46] which also show the updated versions of the mathematical models for the acetone-butanol culture system, the simplest basic update of the batch model appeared first[36].

Other development relevant to the Case Study of this book is centered around different modifications and improvements of the mathematical models describing the acetone-butanol fermentation system. It has to be emphasized that the models of that system presented in this book are by no means perfect and final. They have been constructed as first approximations of the bioreactor system reflecting the current state of knowledge of the process. The model predictions and the exercises using the models, for example bioreactor stability analyses, if confronted with the new and specifically conducted experiments may prove not to reflect the reality quite so accurately. There may be numerous reasons for this: incomplete mass balances, simplistic input of the culture physiology, limited experimental data base or unreliable estimate of the rate expressions, among others. Particularly the latter aspect, reflecting our vague knowledge of the culture kinetics, is a weak point. Much more work remains to be done in carrying out specific experiments before the rate expressions in the mathematical models of the culture systems are reflecting the reality reasonably well. This highlights the first and extremely important feature associated with mathematical modeling of culture systems:

- **the biosystem models can very effectively reduce and**
   meaningfully guide experimental work.

This way, the experimental strategy and methodology can advance from the medieaval "alchemistic" stage of being based on curiosity and looking for answers to randomly generated questions to the stage of experimentation driven by specific questions formulated in an intelligent logical sequence of importance and aimed at elucidating the key elements of the process. The questions can be derived from preliminary experiments simulated on a computer using the mathematical model of the process. That leads to another important power derived from a good process mathematical model:

- **the biosystem models can predict the biosystem behavior and**
  can be used for computer simulation of experiments.

This will enable one to carry out a great number of otherwise tedious and possibly expensive experiments through approximative computer simulations and from a whole set of experiments select only those key and most revealing ones for actual laboratory investigation. In turn, the results of the new set of selected experiments are used for correcting and improving the mathematical model, improving thus its performance for the next round of more acurate simulations.

As demonstrated in the Case Study herein, the mathematical model of the process system can be used to evaluate the system behavior, stability, and sensitivity to changes in specific parameters through the use of routine procedures demonstrated in the Case Study (stability analysis, sensitivity analysis). This approach is particularly possible with mathematical models which are constructed from equations representing mass balances of the relevant chemical process species. These equations have a real physical process basis where each mathematical term reflects a certain aspect or reaction of the process. As a most important consequence

- **the mathematical biosystem model facilitates the system**
  analysis exercises[54],

and   - **allows the process diagnosis**[55].

The availability of reasonably good bioprocess models and their predictive capacity makes it possible to eventually develop the optimal control algorithms for the process which would allow finding the optimal process path and

- **to apply on-line dynamic process control**

during the experimental or, even more desirably, the production run to achieve the highest production efficiencies and productivities. This use of mathematical models of the bioprocess leads into an entirely different and highly involved field of process control.

These possibilities associated with the development and availability of good mathematical models of bioprocess systems are extremely exciting indeed. They open new horizons for studying the bioprocess and its eventual optimization which is in most cases the ultimate goal.

REFERENCES – PART III

1. Bard, Y. 1974. *Nonlinear Parameter Estimation*, Acad. Press, N.Y.
2. Bu'Lock, J.D., Bu'Lock, A.J. Eds. 1983. *The Acetone-Butanol Fermentation and Related Topics*, Sci. Tech. Lett., Kew, Surrey, U.K.
3. de Boor, C. 1978. *A Practical Guide to Splines*, Springer Verlag, N.Y., 1978.
4. Dittmar, R., Hartmann, K., Ostrovsky, G.M. 1978. Theor. Found. Chem. Tech. **12**: 104.
5. Edwards, M.F., Richardson, J.F. 1968. Chem. Eng. Sci. **23**: 109.
6. Finlayson, B. 1972. *The Method of Weighted Residuals and Variational Principles.*, Academic Press, N.Y.
7. Fredrickson, A.G., Ramkrishna, D., Tsuchiya, H.M. 1967. Math. Biosci. **1**: 327.
8. Haggström, L. 1979. in *Proceedings of the 4th Symposium on Technische Mikrobiologie (Berlin)*, Dellweg, H. Ed. p. 271.
9. Haggström, L., Enfors, S.O. 1982. Appl. Biochem. Biotechnol. **7**: 35.
10. Haggström, L., Molin, N. 1980. Biotechnol. Lett. **2**: 35.
11. Hlavacek, V., Votruba, J. 1977. in *Chemical Reaction Theory*, Lapidus, L., Amundsen, N.R. Eds. Prentice Hall, Englewood Cliffs, N.J. p. 314.
12. Harder, A., Roels, J.A. 1982. Adv. Biochem. Eng. **21**: 56.
13. Himmelblau, D.H. 1968. *Process Analysis by Statistical Methods.* J. Wiley, N.Y. p. 114.
14. Kuester, J.L., Mize, J.H. 1973. *Optimization Techniques with Fortran.* McGraw - Hill, N.Y.
15. Leung, J.C.-Y. 1982. Ph.D. thesis, M.I.T., Cambridge, Massachussetts.
16. Malek, I. 1976. in *Continuous Culture 6: Application and new field.* Dean, A.C.R., Ellwood, D.C., Evans, C.G.T., Melling, J. Eds. Ellis Horwood, Chichester, U.K. p. 31.
17. Monot, F., Martin, J.R., Petitdemange, H., Gay, R. 1982. Appl. Environ. Microbiol. **44**: 1318.
18. Moreira, A.R., Ulmer, D.C., Linden, J.C. 1981. Biotech. Bioeng. Symp. **11**: 567.
19. Mulchandani, A. 1985. Ph.D. thesis, McGill University, Montreal, Quebec, Canada.
20. Mulchandani, A., Volesky, B. 1986. Can. J. Chem. Eng., **64**: 625.
21. Naur, P. 1960. Comm. ACM **3**: 312.
22. Perlmutter, D.D. 1972. *Stability of Chemical Reactors.* Prentice Hall, Englewood Cliffs, N.J.
23. Powell, O.E. 1969. in *Proc. 4th Symp. Cont. Cult. Microorg.*, Malek, I. Ed. Publ. House Academia, Prague. p. 275.
24. Ramkrishna, D. 1979. Adv. Biochem. Eng. **11**: 1.
25. Rao, G. 1987. Ph.D. thesis, Drexel University, Philadelphia, PA.
26. Rao, G., Mutharasan, R. 1987. Appl. Environ. Microbiol. **53**: 1232.
27. Rao, G., Mutharasan, R. 1989. Appl. Microbiol. Biotechnol. **30**: 59.
28. Roels, J.A. 1980. Biotech. Bioeng. **22**: 2457.
29. Rosenbrock, H.H. 1960. Computer J., **3**: 175.
30. Schmidt, A. 1983. Acta Biotechnologica **3**: 171.
31. Spivey, M.J. 1978. Process Biochem. **13**(11): 2.

32. Srinivas, S.P., Mutharasan, R. 1987. Biotechnol. Lett. **9**: 139.
33. Srinivas, S.P., Mutharasan, R. 1987. Biotechnol. Bioeng. **30**: 769.
34. Srivastava, A.K. 1990. Ph.D. thesis, McGill University, Montreal, Canada.
35. Srivastava, A.K., Volesky, B. 1988. in *Abstracts 8th Internat. Biotechnol. Symp.*. Paris, France. p. 302.
36. Srivastava, A.K., Volesky, B. 1990. Biotechnol. Lett. 693.
37. Srivastava, A.K., Volesky, B. 1991. Biotechnol. Progress (in press).
38. Srivastava, A.K., Volesky, B. 1991. Appl. Microbiol. Biotechnol. **34**: 450.
39. Srivastava, A.K., Volesky, B. 1991. Biotechnol. Bioeng. **38**: 181.
40. Srivastava, A.K., Volesky, B. 1991. Can. J. Chem. Eng. **69**: 520.
41. Srivastava, A.K., Volesky, B. 1991. Biotechnol. Bioeng. **38**: 191.
42. Takano, M., Tsuchido, T. 1982. J. Ferm. Technol. **60**: 189.
43. Takamatsu, T., Shioya, S., Okuda, K. 1981. J. Ferm. Tech. **59**: 131.
44. Veliky, I. 1987. Personal communication, Division of Biological Sciences, National Research Council of Canada, Ottawa, Ontario.
45. Villadsen, J., Mickelsen, M.L. 1979. *Solution of Differential Equation Models by Polynomial Approximation.* Prentice Hall, Englewood Cliffs, N.J.
46. Volesky, B., Srivastava. A.K. 1991. Anal. Chim. Acta **249**: 279.
47. Volesky B., Votruba, J. 1984. in *Abstracts 7th Int. Biotech. Symp., New Delhi,* **1**: 578.
48. von Bertalanfy, L. 1960. *Fundamental aspects of normal and malignant growth,* Norinski, W.W. Ed. Elsevier, Amsterdam, p. 137.
49. Votruba, J. 1983. *BIOKIN-78 User Manual.* Archive of Computer Programs, Central Computer Centre, Czech Academy of Sciences, Prague, Czechoslovakia.
50. Votruba, J. 1982. Acta Biotechnologica **2**: 119.
51. Votruba, J., Volesky, B. 1984. in *3rd Europ. Biotech. Congress*, Munchen, Verlag Chemie, Weinheim, GFR, **2**: 301.
52. Yerushalmi, L. 1985. Ph.D. thesis, McGill University, Montreal, Canada.
53. Yerushalmi, L., Volesky, B., Leung, W.K. 1983. Eur. J. Appl. Microbiol. Biotechnol. **18**: 279.
54. Yerushalmi, L., Volesky, B., Votruba, J. 1986. Biotechnol. Bioeng. **28**: 1334.
55. Yerushalmi, L., Volesky, B., Votruba, J. 1988. Appl. Microbiol. Biotechnol. **29**: 186.

TABLES OF COMPUTER PROGRAM LISTINGS

TABLE 7.1

LISTING OF COMPUTER PROGRAM BIOKIN-A
FOR DATA SMOOTHING, NUMERICAL DIFFERENTIATION AND
CALCULATION OF THE SUB–PROCESS RATES DURING FERMENTATION

```
/LIST
C      PRELIMINARY DATA ANALYSIS BY SMOOTHING SPLINES: BIOKIN-A
C
       DIMENSION LL(30),YP(40,8),T(40),DY(40),V(40,7),A(40,7),
       *YM(40,8),YD(40,8),Y(40)
50     READ(5,* ) NPOINT,NPROF
       IF(NPOINT.LT.1) STOP
       READ(5,99) (LL(I),I=1,30)
99     FORMAT(30A2)
       WRITE(6,100) (LL(I),I=1,30),NPOINT,NPROF
100    FORMAT(//(30A2/5X,7HNPOINT=,I3,2X,6HNPROF=,I3//)
       DO 1 I=1,NPOINT
       READ(5,*) T(I),(YP(I,J),J=1,NPROF)
1      WRITE(6,101) T(I),(YP(I,J),J=1,NPROF)
101    FORMAT(1X,F4.1,8F7.3)
       S=SQRT(2.*FLOAT(NPOINT))
       DO 10 J=1,NPROF
       DO 11 I=1,NPOINT
       Y(I)=YP(I,J)
11     DY(I)=0.1
       SFS=SMOOTH(T,Y,DY,NPOINT,S,V,A)
       NPM1=NPOINT-1
       DO 12 I=1,NPM1
       YM(I,J)=A(I,1)
12     YD(I,J)=I4(I,2)
       DX=T(NPOINT)-T(NPM1)
       YM(NPOINT,J)=A(NPM1,1)+A(NPM1,2)*DX+A(NPM1,3)*DX**2/2
       *+A(NPM14)*DX**3/6
       YD(NPOINT,J)=A(NPM1,2)+A(NPM1,3)*DX+A(NPM1,4)*DX**2/2
10     CONTINUE
       CALL TRANSF(NPOINT,T,YM,YD)
       GO TO 50
       END
       SUBROUTINE TRANSF(N,X,Y,YD)
       DIMENSION X(40),Y(40,8),YD(40,8),F(20)
       WRITE (6,100)
100    FORMAT(~SX,' TABLE OF RESULTS '/)
       DO 1 I=1,N
       F(1)=YD(I,1)/Y(I,1)
       F(2)=-YD(I,2)/Y(I.1)
       F(3)=YD(I,3)/Y(I,1)
       F(4)=YD(I,4)/Y(I,1)
       F(5)=YD(I,5)/Y(I,1)
       F(6)=YD(I,6)/Y(I,1)
       F(7)=YD(I,7)/Y(I,1)
1      WRITE(6,101) X(1),(F(J),J=1,7)
101    FORMAT(1X,F4.1,7F9.5)
       RETURN
       END
```

(a/continued...b)

TABLE 7.1 (...continued/b)

LISTING OF COMPUTER PROGRAM BIOKIN-A
FOR DATA SMOOTHING, NUMERICAL DIFFERENTIATION AND
CALCULATION OF THE SUB–PROCESS RATES DURING FERMENTATION

```
      REAL FUNCTION SMOOTH (X,Y,DY,NPOINT,S,V,A)
      DIMENSION A(40,4),DY(40),V(40,7),X(40),Y(40)
      CALL SETUPQ(X,DY,Y,NPOINT,V,A(1,4))
      IF(S.GT.O.) GO TO 20
10    P=1
      CALL CHOLID(P,V,A(1,4),NPOINT,1,A(1,3),A(1,1))
      SFP=0.0
      GO TO 60
20     P=O
      CALL CHOLID(P,V,A(1,4),NPOINT,1,A(1,3),A( 1,1))
      SFP=O
      DO 21 I=1,NPOINT
21    SFP=SFP+(A(I,1)*DY(I))**2
      SFP=SFP*36
      IF(SFP.LE.S) GO TO 60
      PREVP=O
      PREVSF=SFP
      UTRU=O
      DO 25 I=2,NPOINT
25    UTRU=UTRU+V(I-1,4)*A(I-1,3)*A(I-1,3)+A(I,3)+A(I,3)**2)
      P=(SFP-S)/(24.*UTRU)
30    CALL CHOLID(P,V,A(1,4),NPOINT,1,A(1,3),A(1,1)
      SFP=O
      DO 35 I=1,NPOINT
35    SFP=SFP+(A(I,1)*DY(I))**2
      SFP=SFP*36.*(1.-P)**2
      IF(SFP.LE.1.O1*S) GO TO 60
      IF(SFP.GE.PREVSF) GO TO 10
      CHANGE=(P-PREVP)/(SFP-PREVSF)*(SFP-S)
      PREVP=P
      P=P-CHANGE
      PREVSF=SFP
      IF(P.LT.1) GO TO 30
      P=1.-SQRT(S/PREVSF)*(1.-PREVP)
      GO TO 30
60    SMOOTH=SFP
      SIX1MP=6.*(1.-P)
      DO 61 I=1,NPOINT
61    A(I,1)=Y(I)-SIX1MP*DY(I)**2*A(I,1)
      SIXP=6.*P
      DO 62 I=1,NPOINT
62    A(I,3)=A(I,3)*SIXP
      NPM1=NPOINT-1
      DO 63 I=1,NPM1
      A(I,4)=(A(I+1,3)-A(I,3))/V(I,4)
63    A(I,2)=(A(I+1,1)-A(I,1))/V(I,4)
     *-(A(I,3)+A(I,4)/3.*V(I,4)/2.*V(I,4)
      RETURN
      END
```

(b/continued...c)

TABLE 7.1 (...continued/c)

**LISTING OF COMPUTER PROGRAM BIOKIN-A
FOR DATA SMOOTHING, NUMERICAL DIFFERENTIATION AND
CALCULATION OF THE SUB–PROCESS RATES DURING FERMENTATION**

```
      SUBROUTINE CHOLID(P,V,QTY,NPOINT,NCOL,U,QU)
      DIMENSION QTY(40),QU(40),U(40),V(40,7)
      NPM1=NPOINT-1
      SIX1MP=6.*(1.-P)
      TWOP=2.*P
      DO 2 I=2,NPM1
      V(I,1)=SIX1MP*V(I.5)+TWOP*(V(I-1,4)+V(I,4)
      V(I,2)=SIX1MP*V(I,6) +P*V(I,4)
2     V(I,3)=SIX1MP*V(I,7)
      NPM2=NPOINT-2
      IF(NPM2.GE.2) GO TO 10
      U(1)=O.
      U(2)=QTY(2)/V(2,1)
      U(3)=0.O
      GO TO 41
10    DO 20 I =2,NPM2
      RATIO=V(I,2)/V(I,1)
      V(I+1,1)=V(I+1,1)-RATIO*V(I,2)
      V(I+1,2)=V(I+1,2)-RATIO*V(I,3)
      V(I,2)=RATIO
      RATIO=V(I,3)/V(I,1)
      V(I+2,1)=V(I+2,1)-RATIO*V(I,3)
20    V(I,3)=RATIO
      U(1)=0
      V(1,3)=0.0
      U(2)=QTY(2)
      DO 30 I=2,NPM2
30    U(I+1)=QTY(I+1)-V(I,2)*U(I)-V(I-1,3)*U(I-1)
      U(NPOINT)=0.0
      U(NPM1)=U(NPM1)/V(NPM1,1)
      DO 40 II=2,NPM2
      I=NPOINT-II
40    U(I)=U(I)/V(I,1)-U(I+1)*V(I,2)-U(I+2)*V(I,3)
41    PREV=0.0
      DO 5O I=2,NPOINT
      QU(I)=(U(I)-U(I-1))/V(I-1,4)
      QU(I-1)=QU(I)-PREV
50    PREV=QU(I)
      QU(NPOINT)=-QU(NPOINT)
      RETURN
      END
```

(c/continued...d)

TABLE 7.1 (...continued/d)

LISTING OF COMPUTER PROGRAM BIOKIN-A
FOR DATA SMOOTHING, NUMERICAL DIFFERENTIATION AND
CALCULATION OF THE SUB–PROCESS RATES DURING FERMENTATION

```
      SUBROUTINE SETUPQ(X,DX,Y,NPOINT,V,QTY)
      DIMENSION DX(40),QTY(40),V(40.7),X(40),Y(40)
      NPM1=NPOINT-1
      V(1,4)=X(2)-X(1)
      DO 11 I=2,NPM1
      V(I,4)=X(I+1)-X(I)
      V(I,1)=DX(I-1)/V(I-
1,4)
      V(I,2)=-DX(I)/V(I,4)-DX(I)/V(I-
1,4)
      V(I,3)=DX(I+1)/V(I,4)
      V(NPOINT,1)=0.
      DO 12 I=2,NPM1
12    V(I,5)=V(I,1)**2+V(I,2)**2+V(I,3)**2
      IF(NPM1.LT.3) GO TO 14
      DO 13 I=3,NPM1
13    V(I-1,6)=V(I-1,2)*V(I,1)+V(I-1,3)*V(I,2)
14    V(NPM1,6)=0.
      IF(NPM1.LT.4) GO TO 16
      DO 15 I=4,NPM1
15    V(I-2,7)=V(I-2,3)*V(I,1)
16    V(NPM1-1,7)=0.0
      V(NPM1,7)=0.
      PREV=(Y(2)-Y(1))/V(1,4)
      DO 21 I=2,NPM1
      DIFF=(Y(I+1)-Y(I))/V(I,4)
      QTY(I)=DIFF-PREV
21    PREV=DIFF
      RETURN
      END
```

TABLE 7.2

APPLICATION EXAMPLE
OF COMPUTER PROGRAM BIOKIN-A

a) INPUT LISTING

```
/DATA
THE A-B-E FERMENTATION PROCESS: DATA BATCH CULTIVATION NO.7
    0.0      0.06    52.0     0.0      0.0      0.0      0.0      0.0
    3.0      0.06    52.0     0.0      0.0      0.0      0.0      0.0
    8.0      0.2O    50.5     0.1      0.05     0.0      0.3      0.3
   12.0      0.49    44.0    15 0      0.1      0.02     1.5      1.5
   15.0      1.75    38.      0.5      0.25     0.05     2.3      2.5
   17.5      3.5     29.5     1.0      0.45     0.1      2.5      2.9
   21.0      3.9     19.5     4.75     2.1      0.42     0.5      1.9
   24.0      3.85     8.0     9.3      4.75     0.96     0.4      1.9
   27.0      3.4      0.5    10.9      5.1      1.0      0.2      1.9
   32.5      3.25     0.0    10.9      5.1      1.0      0.2      1.9
0   0
/ENDRUN
*END
```

b) RESULT PRINT-OUT

```
                                    /EXEC JV
                                    *IN PROGRESS
                                    MAIN  = 002026
                                    TRANSF = 0002D4
                                    SMOOTH = 0006CO
                                    SETUPQ = 0003EE
                                    CHOLID = OOO5OE
                                    0077FO BYTES USED
                              EXECUTION BEGINS
THE A-B-E FERMENTATION PROCESS:DATA BATCH CULTIVATION NO.7
     NPOINT=10        NPROF=7
    0.0      0.060   52.000    0.0      0.0      0.0      0.0      0.0
    3.0      0.060   52.000    0.0      0.0      0.0      0.0      0.0
    8.0      0.200   50.500    0.100    0.050    0.0      0.300    0.300
   12.0      0.490   44.000    0.150    0.100    0.020    1.500    1.500
   15.0      1.750   38.000    0.500    0.250    0.050    2.300    2.500
   17.5      3.500   29.500    1.000    0.450    0.100    2.500    2.900
   21.0      3.900   19.500    4.750    2.100    0.420    0.500    1.900
   24.0      3.850    8.000    9.300    4.750    0.960    0.400    1 900
   27.0      3.400    0.500   10.900    5.100    1.000    0.200    1 900
   32.5      3.250    0.0     10.900    5.100    1.000    0.200    1 900

TABLE OF RESULTS:
    0.0     -0.0290   0.1557  -0.0511  -0.0204  -0.0373  -0.0521  -0.1195
    3.0      0.1779  -0.4974   0.1660   0.0677  -0.0373  -0.0264   0.0221
    8.0      0.0869   6.4046   0.0521   0.0572  -0.0053   1.1946   1.1704
   12.0      0.4912   3.3004   0.0996   0.0496   0.0200   0.6206   0.6743
   15.0      0.3256   1.5332   0.0615   0.0237   0.0189   0.1087   0.1319
   17.5      0.1350   0.9534   0.1673   0.0671   0.0205  -0.0851  -0.0216
   21.0      0.0026   0.8S98   0.3738   0.1986   0.0260  -0.0899  -0.0440
   24.0     -0.0274   0.9175   0.2870   0.1413   0.0219  -0.0020   0.0008
   27.0     -0.0321   0.4242   0.0712   0.0030   0.0110  -0.0120   0.0029
   32.5      0.0010  -0.1791  -0.0397  -0.0067   0.0002   0.0051  -0.0008
STOP               0
```

TABLE 7.3

COMPUTER PROGRAM BIOKIN-B
PRINT-OUT OF THE INPUT MODULE

```
        DIMENSION
YP(20,8),T(20),A(20),EPS(20),LL(30),W(10)
        COMMON /DAT/ NPOINT,NPROF,T,W,YP
        READ(5,*)NPOINT,NPROF,NPAR,MFC
        READ(5,99) (LL(I),I=1,30)
99      FORMAT(30A2)
        WRITE(6,100) (LL(I),I=1,30)
100     FORMAT(//1X,30A2)
        DO 1 I=1,NPOINT
        READ(5,*) T(I),(YP(I,J),J=1,NPROF)
1       WRITE(6,101) T(I)(YP(I.J),J=1,NPROF)
101     FORMAT(1X,F4.1,8F7.3)
        READ(5,*) (A(I),I=1,NPAR)
        READ(5,*)(EPS(I),I=1,NPAR)
        READ(5,*) (W(I),I=1,NPROF)
        CALL ROS(NPAR,MFC,A,EPS,0.001)
        STOP
        END


        SUBROUTINE OBJECT(A,SSWR,NPAR)
        DIMENSION
AA(20),A(20),W(10),YP(20,8),T(20),Y(20)
        COMMON /KON/AA
        COMMON /DAT/ NPOINT,NPROF,T,W,YP
        WRITE(6,120) (AI),I=1,NPAR)
120     FORMAT(1X,10F7.4)
        DO 5 I=1,NPAR
5       AA(I)=ABS(A(I))
        SSWR=0.0
        NPM1=NPOINT-1
        DO 1 I-1,NPM1
        IF(I.GT.1) GO TO 3
        DO 2 J=1,NPROF
2       Y(J)=YP(1,J)
3       XB=T(I)
        XF=T(I+1)
        CALL ODE(NPROF,XB,YF,Y,0.01)
        DO 4 J=1,NPROF
4       SSWR=SSWR+((YP(I+1,J)-Y(J))/W(J))**2
        CONTINUE
        WRITE(6,110) SSWR
110     FORMAT(1X,E12.5)
        RETURN
        END
```

(a/continued...b)

TABLE 7.3 (...continued/b)

COMPUTER PROGRAM BIOKIN-B
PRINT-OUT OF THE INPUT MODULE

```
      SUBROUTINE RHS(N,X,Y,F)
      DIMENSION Y(20),F(20),A(20),AA(20)
      COMMON /KON/ AA
      A(1)=0.63
      A(2)=0.4
      A(3)=0.7
      A(4)=0.024
      A(5)=0.0107
      DO 1 I=1,5
1     A(I+5)=AA(I)
      FS=Y(2)/(Y(2)+A(9))
      FI=A(1)/(A(1)+Y(3))
      F(1)=A(2)*0.02*Y(2)*FI*Y(1)-0.0014*Y(1)*Y(3)
      F(2)=-A(3)*FS*Y(1)-A(4)*Y(2)*Y(1)
      F(6)=(A(10)*Y(2)*FI-0.1*Y(6)/(Y(6)+0.5))*Y(1)
      F(3)=A(5)*Y(2)*Y(1)-0.831*F(6)
      F(7)=(A(6)*FS*FI-0.01*Y(7)/(Y(7)+0.5))*Y(1)
      F(4)=A(7)*FS*Y(1)-F(7)*0.498
      F(5)=A(8)*Y(1)*FS
      RETURN
      END
/INCLUDE JERRYOPT
/INCLUDE JERRYODE
/DATA
10 7 5 30
DATA BATCH CULTIVATION  NO 7:THE A-B-E FERMENTATION
     0.0       0.03      52.0      0.0       0.0       0.0       0.0       0.0
     3.0       0.06      52.0      0 0       0 0.      0 0       0 0       0.0
     8.0       0.20      50.5      0.1       0.05      0.0       0.3       0.3
    12.0       0.49      44.0      0.15      0.1       0.02      1.5       1.5
    15.0       1.75      38.0      0.5       0.25      0.05      2.3       2.5
    17.5       3.5       29.5      1.0       0.45      0.1       2.5       2.9
    21.0       3.9       19.5      4.75      2.1       0.42      0.5       1.9
    24.0       3.85      8.0       9.3       4.75      0.96      0.4       1.9
    27.0       3.4       0.5       10.9      5.1       1.0       0.2       1.9
    32.5       3.25      0.0       10.9      5 1       1.0       0.2       1.9
     0.3       0.132     0.023     1.0       0.0135
     0.03      0.04      0.003     0.5       0.004
     4.0      40.0       8.0       4.0       1.0       2.0       2.0
     0.0       0.0       0.0       0.0       0.0       0.0
/ENDRUN
*END
```

TABLE 7.4

LISTING OF THE INTEGRATION COMPUTER PROGRAM
FOR SOLVING SYSTEMS OF ORDINARY DIFFERENTIAL EQUATIONS
BY THE RUNGE-KUTTA METHOD

```
      SUBROUTINE ODE (N,XB,XE,Y,EPS)
      DIMENSION Y1(20),Y2(20),Y3(20),Y(20)
      X=XB
      H=(XE-X)/10.
C     ADOPTED FROM ALGORITHM 8 C ACM(1960) VOL.3, P.312
      IS=O
      IOUT=O
100   IF((X+2.01*H-XE).LT.0) GO TO 1
      IOUT=1
      H=(XE-X)/2.
1     CALL RK4(N,X,Y,2.*H,X1,Y1)
101   CALL RK4(N,X,Y,2,H,X2,Y2)
      CALL RK4(N,X2,Y2,H,X3,Y3)
      ERR=0.0
      DO 2 K=1,N
      Q1=AMAX1(1.E-6,ABS(Y3(K)))
      P=ABS(Y1(K)-Y3(K))/Q1
2     ERR=AMAX1(P,ERR)
      IF(ERR.GT.EPS) GO TO 103
      X=X3
      IF(IOUT.EQ1) GO TO 104
      DO 3 K=1,N
3     Y(K)=Y3(K)
      IF(IS.LT.5) GO TO 4
      H=2 *H
      IS=0
4     IS=IS+1
      GO TO 100
103   H=0.5*H
      IOUT=O
      X1=X2
      DO 5 K=1,N
5     Y1(K)=Y2(K)
      GO TO 101
104   DO 6 K=1,N
6     Y(K)=Y3(K)
      RETURN
      END
      SUBROUTINE RK4 (N,X,Y,H,XE,YE)
      DIMENSION Y(20),YE(20),Z(20),W(20),A(5)
      DATA A /0.5,0.5,1.,1.0,0.5/
      XE=X
      DO 1 K=1,N
      YE(K)=Y(K)
1     W(K)=Y(K)
      DO 2 J=1,4
      CALL RHS(N,XE,W,Z)
      XE=A(J)*H+X
      DO 3 K=1,N
      W(K)=Y(K)+A(J)*H*Z(K)
3     EE(K)=YE(K)+A(J+1)*H*Z(K)/3
2     CONTINUE
      RETURN
      END
```

TABLE 7.5

LISTING OF THE COMPUTER PROGRAM FOR NUMERICAL MINIMIZATION

```
       SUBROUTINE ROS  (KM,MAXK,AFK,EPS,EPSY)
       DIMENSION AKE(20),D(20),V(20,20),BL(20,20),BLEN(20),EPS(20),
           1AJ(20), E(20),AL(2O,2O),AFK(20)
       NO=6
       NSTEP=2
       MKAT=30
       MCYC=50
       ALPHA=2.0
       BETA=0.5
       WRITE(NO,99)
99     FORMAT(1H1,10X,34HROSENEBROCK MINIMIZATION PROCEDURE)
       WRITE(NO,1004)  MAXK,MKAT,MCYC,NSTEP,ALPHA,BETA,EPSY
1004   FORMAT(//2X,1OHPARAMETERS/2X,7HMAXK=,I4,4X,7HMKAT=,I2,4X,
       1 7HMCYC = I2,4X,8HNSTEP = ,I2//2X,8HALPHA = ,F5.2,4X,
       2 7HBETA = ,F5.2,4X,7HEPSY = ,1E12.4)
       KAT=1
       DO 98 II=1,KM
       DO 98 JJ=1,KM
       V(II,JJ)=0.0
       IF(II-JJ) 98,97,98
97     V(II,JJ)=1
98     CONTINUE
       CALL OBJECT(AFK,SUMN,KM)
       SUMO=SUMN
       DO 812 K=1,KM
       AKE(K)=AFK(K)
812    CONTINUE
       KK1=1
       IF(NSTEP-1) 701,700,701
700    GO TO 1000
701    CONTINUE
       DO 350 I=1,KM
       E(I)=EPS(I)
350    CONTINUE
1000   DO 250 I=1,KM
       FBEST=SUMN
       AJ(I)=2
       IF(NSTEP-1) 702,703,702
702    GO TO 250
703    CONTINUE
       E(I)=EPS(I)
250    D(I)=0.0
       III=0
397    III=III+1
258    I=1
259    DO 251 J=1,KM
251    AKE(J)=AKE(J)+E(I)*V(I,J)
       CALL OBJECT(AKE,SUMN,KM)
       KAT=KAT+1
       IF(KAT-MAXK) 706,707,707
707    GO TO 1001
706    CONTINUE
       IF(SUMN-SUMO) 708,708,709
708    GO TO 253
709    CONTINUE                          (a/continued...b)
```

TABLE 7.5 (...continued/b)

LISTING OF THE COMPUTER PROGRAM FOR NUMERICAL MINIMIZATION

```
         DO 254 J=1,KM
254      AKE(J)=AKE(J)-E(I)*V(I,J)
         E(I)=-BETA*E(I)
         IF(AJ(I)-1.5)710,711,711
710      AJ(I)=0.0
711      CONTINUE
         GO TO 255
253      D(I)=D(I)+E(I)
         E(I)=ALPHA*E(I)
         SOMO=SUMN
         DO 813 K=1,KM
813      AFK(K)=AKE(K)
         IF(AJ(I) -1.5) 712,712,713
713      AJ(I)=1.0
712      CONTINUE
255      DO 256 J=1,KM
         IF(AJ(J)-0.5) 256,256,715
715      GQ TO 299
256      SUMDIF=FBEST-SUMO
         SUMDIF=SUMDIF/AMAX1(ABS(FBEST),0.001)
         IF(ABS(SUMDIF)-EPSY)1001,1001,257
299      IF(I-KM) 717,716,717
716      GO TO 399
717      CONTINUE
         I=I+1
         GO TO 259
399      DO 398 J=1,KM
         IF(AJ(J)-2.) 718,398,398
718      GO TO 258
398      CONTINUE
         IF(III-MCYC) 720,721,721
720      GO TO 397
721      CONTINUE
         GO TO 1001
257      CONTINUE
         DO 290 I=1,KM
         DO 290 J=1,KM
290      AL(I,J)=0.0
         WRITE(NO,280) KK1
280      FORMAT(//2X,13HSTAGE NUMBER ,I2)
         WRITE(NO,281) SUMO
281      FORMAT(/7X,31HVALUE OF OBJECTIVE FUNCTION = ,E12.5)
         WRITE(NO,282)
282      FORMAT(/7X,35HVALUES OF THE INDEPENDENT VARIABLES /)
         DO 284 IX=1,KM
         WRITE(NO,283) IX,AKE(IX)
283      FORMAT(/7X,2HX(I2,4H) = ,E16.8)
284      CONTINUE
         DO 280 I=1,KM
         KL=I
         DO 260 J=1,KM
         DO 261 K=KL,KM
261      AL(I,J)=D(K)*V(K,J)  + AL(I,J)
260      BL(I,J) = AL(I,J)                        (b/continued...c)
```

TABLE 7.5 (...continued/c)

LISTING OF THE COMPUTER PROGRAM FOR NUMERICAL MINIMIZATION

```
          BLEN(1)=0.0
          DO 351 K=1,KM
          BLEN(1)=BLEN(1) +BL(1,K)*BL(1,K)
351       CONTINUE
          BLEN(1)=SQRT(BLEN(1))
          DO 352 J=1,KM
          V(1,J)=BL(1,J)/BLEN(1)
352       CONTINUE
          DO 263 I=2,KM
          II=I-1
          DO 263 J=1,KM
          SUMAVV=0.0
          DO 264 KK=1,II
          SUMAV=0.0
          DO 262 K=1,KM

262       SUMAV = SUMAV+ AL(I,K)*V(KK,K)
264       SUMAVV=SUMAV*V(KK,J)+SUMAVV
263       BL(I,J)=AL(I,J) -SUMAVV
          DO 266 I=2,KM
          BLEN(I)=0.0
          DO 267 K=1,KM
267       BLEN(I)=BLEN(I) + BL(I,K)*BL(I,K)
          BLEN(I)=SQRT(BLEN(I))
          DO 266 J=1,KM
266       V(I,J)=BL(I,J)/BLEN(I)
          KK1=KK1+1
          IF(KK1-MKAT) 723,722,722
722       GO TO 1001
723       GO TO 1000
1001      WRITE(NO,1002) KK1
1002      FORMAT(///2X,25HTOTAL NUMBER OF STAGES = ,I2)
          WRITE(NO,1003) KAT
1003      FORMAT(/2X,38HTOTAL NUMBER OF FUNCTION EVALUATION ,I5)
          WRITE( NO,1005) SUMO
1005      FORMAT(/2X,33HFINAL.VALUE OF OBJECT FUNCTION = ,E12.5)
          DO 1007 IX=1,KM
          WRITE(NO,1006) IX,AKE(IX)
1006      FORMAT(/2X,2HX(,I2,4H) = ,E16.8)
1007      CONTINUE
          RETURN
          END
```

TABLE 7.6

EXAMPLE OF A BIOKIN-B PROGRAM RUN

```
MAIN   = OOO504
OBJECT = OOO3A4
RHS    = OOO35A
ROS    = OO2498
ODE    = OOO584
RK4    = OOO364
OO83OO BYTES USED
EXECUTION BEGINS
```

DATA BATCH CULTIVATION NO 7: THE A-B-E FERMENTATION PROCESS

0.0	0.030	52.000	0.0	0.0	0.0	0.0	0.0
3.0	0.060	52.000	0.0	0.0	0.0	0.0	0.0
8.0	0.200	50.000	0.100	0.050	0.0	0.300	0.300
12.0	0.490	44.000	0.150	0.100	0.020	1.SOO	1.500
15.0	1.750	38.000	0.500	0.250	0.050	2.300	2.500
17.5	3.500	29.500	1.000	0.450	0.100	2.500	2.900
21.0	3.900	19.500	4.750	2.100	0.420	0.~00	1.900
24.0	3.850	8.000	9.300	4.7SO	0.~60	0.400	1.~00
27.0	3.400	0.500	10.900	5.100	1.000	0.200	1.900
32.5	3.250	0.0	10.900	5.100	1.000	0.200	1.900

ROSENBROCK MINIMIZATION PROCEDURE

PARAMETERS

MAXK = 30   MKAT = 30   MCYC = 50   NSTEP - 2

ALPHA = 2.00   BETA = 0.50   EPSY = 0.1OOOE-02

```
     0.3000        0.1320        0.0235        1.0000        0.0135
0.423S2E+O1
     0.3300        0.1320        0.0235        1.0000        0.0135
0.43418E+O1
     0.3000        0.1720        0.0235        1.0000        0.0135
0.44603E+O1
     0.3000        0.1320        0.0265        1.0000        0.0135
0.43407E+O1
     0.3000        0.1320        0.0235        1.5000        0.0135
0.42565E+O1
     0.3000        0.1320        0.0235        1.0000        0.0175
 *TIME ESTIATE OF 8 SERVICE UNITS EXCEEDED---JOB TERMINATED
 *END
 *GO
```

TABLE 7.9

LISTING OF THE COMPUTER PROGRAM FOR
PARAMETRIC SENSITIVITY CALCULATION

```
C    PARAMETRIC SENSITIVITY OF THE A-B-E FERMENTATION PROCESS MODEL

         DIMENSION YP(20,81),T(20),A(20),LL(30),W(10)
         COMMON /DAT/ NPOINT,NPROF,T,W,YP
         READ(5,*) NPOINT,NPROF,NPAR
         READ(5,99) (LL(I),I=1,30)
99       FORMAT(30A2)
         WRITE(6,100) (LL(I),I=1,30)
100      FORMAT(//1X,30A2//)
         DO 1 I=1,NPOINT
         READ(5,*) T(I),(~P(I,J),J-1,NPROF)
101      FORMAT(1X,F4.1,8F7.3)
         READ(5,*) (A(I),I=1,NPAR)
         READ(5,*) (W(I),I=1,NPROF)
         CALL OBJECT(A,SS1,NPAR)
         DO 2 I=1,NPAR
         DA=0.01*A(I)
         A(I)=A(I)+DA
         CALL OBJECT(A,SS2,NPAR)
         APS=(SS2-SS1)/DA
         A(I)=A(I)-DA
         RPS=A(I)*ABS(APS)/SS1
2        WRITE(6,102) A(I),APS,RPS
102      FORMAT(1X,F6.3,2X,2E12.5)
         STOP
         END

         SUBROUTINE OBJECT(A,SSWR,NPAR)
         DIMENSION AA(20),A(20),W(10),YP(~20,8),T(20),Y(20)
         COMMON /KON/ AA
         COMMON /DAT/ NPOINT,NPROF,T,W.YP
         DO 5 I=1,NPAR
5        AA(I)=ABS(A(I))
         SSWR=0.0
         NPM1=NPOINT-1
         DO 1 I=1,NPM1
         IF(I.GT.1) GO TO 3
         DO 2 J=1,NPROF
```

(a/continued...b)

TABLE 7.9 (...continued/b)

LISTING OF THE COMPUTER PROGRAM FOR
PARAMETRIC SENSITIVITY CALCULATION

```
        Y(J)  =  YP(I,J)
3       XB=T(I)
        XF=T(1+1)
        CALL  ODE(NPROF,XB,XF,Y,0.01)
        DO 4 J=1,NPROF
4       SSWR=SSWR+((YP)I+1,J)-Y(J))/W(J))**2
        CONTINUE
        RETURN
        END

        SUBROUTINE RHS  (N,X,Y,F)
        DIMENSION Y(20) ,F(20) ,A(20)
        COMMON /KON/ A
        FS=Y(2)/(Y(2)+A(9))
        FI=A(1)/A(1)+Y(3))
        F(1)=A(2)*0.02*Y(2)*FI*Y(1)-A(11)*Y(1)*Y(3)
        F(2)=-A(3)*FS*Y(1)-A(4)*Y(2)*Y(1)
        F(6)=(A(10)*Y(2)*FI-A(12)*Y(6)/(Y(6)+A(13)))*Y(1)
        F(3)=A(5)*Y(2)*Y(1)-0.831*F(6)
        F(7)=(A(6)*FS*FI-A(15)*Y(7)/(Y(7)+A(14)))*Y(1)
        F(4)=A(7)*FS*Y(1)-F(7)*0.498
        F(5)  =A(8)*Y(1)*FS
        RETURN
        END
/INCLUDE ODE
/DATA
10 7 15
DATA BATCH CULTIVATION NO 7: THE A-B-E FERMENTATION PROCESS
      0.0       0.03      52.0       0.0       0.0       0.0       0.0       0.0
      3.0       0.06       2.0       0.0       0.0       0.0       0.0       0.0
      8.0       0.20      50.5       0.1       0.05      0 0       0.3       0.3
     12.0       0.49      44.0       0.15      0.1       0.02      1.5       1.5
     15.0       1.75      38.0       0.5       0.25      0.05      2.3       2.5
     17.5       3.5       29.5       1.0       0.45      0.1       2.5       2.9
     21.0       3.9       19.5       4.75      2.1       0.42      0.5       1.9
     24.0       3.85       8.0       9.3       4.75      0.96      0.4       1.9
     27.0       3.4        0.5      10.9       5.1       1.0       0.2       1.9
      3 2       3.25       0.0      10.9       5.1       1.0       0.2       1.9
      0.63      0.4        0.7       0.02      0.01      0.3       0.13      0.02
/ENDRUN
*END

*GO
```

TABLE 7.9 (...continued/c)

PARAMETRIC SENSITIVITY ANALYSIS
EXAMPLE OF A RESULT PRINT-OUT

```
MAIN    = 0004E2
OBJECT  = 000320
RHS     = 000276
ODE     = 000584
RK 4    = 000364
05B70 BYTES USED, EXECUTION BEGINS


DATA BATCH CULTIVATION NO 7: THE A-B-E FERMENTATION PROCESS

        0.630       0.75808E+03     0.15714E+01
        0.400       0.67493E+04     0.88826E+01
        0.700       0.78826E+03     0.18155E+01
        0.024       0.26111E+05     0.20618E+01
        0.011      -0.56350E+06     0.19838E+02
        0.300       0.11719E+02     0.11567E-01
        0.132      -0.13261E+03    0 57595E-01
        0.023       0.36361E+02     0.28115E-02
        1.000      -0.16016E+02     0.52695E-01
        0.013       0.83855E+06     0.37247E+02
        0.001      -0.65046E+04     0.29962E-01
        0.100      -0.55832E+04     0.18370E+01
        0.500       0.48730E+03     0.80167E+00
        0.500       0.43945E+00     0.72294E-03
        0.010      -0.10254E+03     0.33737E-02
STOP                     O
*END

*GO
/OFF
*GOOD-BYE
```

TABLE 7.11

**LISTING OF A COMPUTER PROGRAM FOR THE
CONTINUOUS-FLOW CULTURE SYSTEM**

```
C        SIMULATION OF THE A-B-E CONTINUOUS-FLOW FERMENTATION PROCESS

         DIMENSION YP(20,8),T(20),A(20),LL30,W(10)
         COMMON /DAT/ NPOINT,NPROF,T,W,YP
         COMMON/PAR/D,SO
         READ(5,*) NPOINT,NPROF,NPAR
         READ(5,*) T(1),YP(1,J),J=1,NPROF
         FORMAT(1X,F4.1,8F7.3)
         READ(5,*) (A(I),I=1,NPAR)
         READ(5,*) D,SO
         WRITE(6,99) D,SO
99       FORMAT(//2HD=,F6.4,1X,3HSO=,F5.1/)
         CALL OBJECT(A,SS2,NPAR)
2        CONTINUE
         STOP
         END

         SUBROUTINE OBJECT (A,SSWR,NPAR)
         DIMENSION AA(20),A(20),W(10),YP(20,8),T(20),Y(20)
         COMMON /KON/ AA
         COMMON /DAT/ NPOINT,NPROF,T,W,YP
         DO 5 I=J,NPAR
5        AA(I)=ABS(A(I))
         NROV=NPROF+3
         NRM1=NROV-3
         XF=T(1)
         NPM1=NPOINT-1
         DO 1 I=1,NPM1
         IF(I.GT.1) GO TO 3
         DO 2 J=1,NPROF
2        Y(J)=YP(1,J)
         Y(NPROF+1)=0.0
         Y(NPROF+2)=0.0
         Y(NPROF+3)=1.2
3        XB=XF
         XF=XB+5.0
         CALL ODE (NROV,XB,XF,Y,0.01)
         WRITE(6,100) XF,(Y(JJ),JJ=1,NRM1)
100      FORMAT(1X,F4.0,F5.2,F5.1,F6.2,5F5.2,F5.1)
         CONTINUE
         RETURN
         END
```

(a/continued...b)

TABLE 7.11 (...continued/b)

**LISTING OF A COMPUTER PROGRAM FOR THE
CONTINUOUS-FLOW CULTURE SYSTEM**

```
      SUBROUTINE RHS (N,X,Y,F)
      DIMENSION Y(20),F(20),A(20)
      COMMON /KON/ A
      COMMON/PAR/D,SO
      FS=Y(2)/(Y(2)+A(8))
      FI=A(1)/(A(1)+Y(3))
      F(1)=0.56*(Y(10)-1.)*Y(1)-A(9)*Y(1)*Y(3)-D*Y(1)
      FIB=A(7)/A(7)=Y(3)
      F(6)=A(5)*FS*FIB*Y(1)-D*Y(6)
      F(2)=-A(4)*Y(2)*Y(1)-A(3)*FS*Y(1)=D*(SO-Y(2))
      F(3)=A(6)*FS*Y(1)-0.831*(F(6)+D*Y(6)-D*Y(3)
      F(7)=A(11)*FS*FI*Y(1)-D*Y(7)
      F(4)=A(12)*FS*Y(1)-(F(7)+Y(7)*D)*0.498-*Y(4)
      F(5)=A(13)*Y(1)*FS-D*Y(5)
      F(9)=0.6*FS*Y(1)
      F(8)=0.019*FS*Y(1)
      F(10)=(A(2)*FI*FS-0.56*(Y(10)-1.))*Y(10)
      RETURN
      END
/INCLUDE ODE
/DATA
25 7 15
    0.0   20.02   5.0   12.0    0.81   0.24  5.88  3.9   0.0    0.0     1.2
    5.0    0.44   1.25  0.005   0.6    0.367 0.97  1.0   0.00140.0      0.22
    0.187  0.033  0.0   0.0     0.1   50.0
*END
```

TABLE 7.14

LISTING OF A COMPUTER PROGRAM FOR THE CELL-RECYCLE CONTINUOUS-FLOW CULTURE SIMULATION

```
       SUBROUTINE RHS (N,X,Y,F)
       DIMENSION Y(20),F(20),A(20)
       COMMON /KON/A
       COMMON/PAR/D,S0
       FS=Y(2)/(Y(2)+A(B))
       FI=A(1)/(A(1)+Y(3))
       F(1)=0.56*(Y(10)-1.)*Y(1)-A(9)*Y(1)*(3)
       F(2)=-A(3)*FS*Y(1)-A(4)*Y(2)*Y(1)+D*(S0-Y(2))
       F(6)=(A(5)*Y(2)*FI-A(7)*Y(6)/(Y(6)+A(10)))*Y(1)-D*Y(6)
       F(3)=A(6)*Y(2)*Y(1)-0.831*(F(6)+D*Y(6))-D*Y(3)
       F(7)=(A911)*FS*FI-A(15)*Y(7)/(Y(7)+A(14)))*Y(1)-D*Y(7)
       F(4)=A(12)*FS*Y(1)-(F(7)+Y(7)*D)*0.498-D*Y(4)
       F(5)=A(13)*Y(1)*FS-D*Y(5)
       F(9)=0.6*FS*Y(1)
       F(8)=0.019*FS*Y(1)
       F(10)=(A(2)*FI*FS-0.56*(Y(10)-1.))*Y(1)
       RETURN
       END
/INCLUDE JERRYODE
/DATA
15 7 15
0.0  4.24  3.3  1.84  0.81  0.24  5.88  3.9  0.0  0.0  1.2
0.42  0.38  0.85  .03  0.27  .09  .05  1.0  .0014  .5  .66  .132
.018  .5  0.01
0.025
/ENDRUN
*END

GO
/exec Jerryfed
*IN PROGRESS
```

(a/continued...b)

TABLE 7.14(...continued/b)

LISTING OF A COMPUTER PROGRAM FOR THE
CELL-RECYCLE CONTINUOUS-FLOW CULTURE SIMULATION

```
/LOAD FORTG1
C
C       SIMULATION OF THE A-B-E PROCESS:
                         CELL-RECYCLE CONTINUOUS-FLOW CULTURE
        DIMENSION YP(20,8), T(20), A(20), LL(30), W(10)
        COMMON /DAT/ NPOINT,NPROF,T,W,YP
        COMMON/PAR/D,S0
        READ(5,*) NPOINT,NPROF,NPAR
        READ(5,*) T(1),(YP(1,J),J=1,NPROF)
101     FORMAT(1X,F4.1,8F7.3)
        READ(5,*) (A(I),I=1,NPAR)
        READ(5,*) D
        S0=0.0
        DO 2 I=1,5
        S0=S0+10.
        WRITE(6,99) D,S0
99      FORMAT(//2HD=,F6.4,1X,3HS0=,F5.1/)
        CALL OBJECT(A,SS2,NPAR)
2       CONTINUE
        STOP
        END
        SUBROUTINE OBJECT(A,SSWR,NPAR)
        DIMENSION AA(20),A(20),W(10),YP(20,8),T(20),Y(20)
        COMMON /KON/ AA
        COMMON /DAT/ NPOINT,NPROF,T,W,YP
        DO 5 I=1,NPAR
5       AA(I)=ABS(A(I))
        NROV=NPROF+3
        NRM1=NROV-1
        XF=T(1)
        NPM1=NPOINT-1
        DO 1 I=1,NPM1
        IF(I.GT.1) GO TO 3
        DO 2 J=1,NPROF
2       Y(J)=YP(1,J)
        Y(NPROF+1)=0.0
        Y(NPROF+2)=0.0
        Y(NPROF+3)=1.2
3       XB=XF
        XF=XB+5.0
        CALL ODE(NROV,XB,XF,Y,0.01)
        WRITE(6,100) XF,(Y(JJ),JJ=1,NRM1)
100     FORMAT(1X,F3.0,F5.2,F5.1,F6.2,F5.2,F5.1)
        CONTINUE
        RETURN
        END
```

TABLE 7.17

COMPUTER PROGRAM LISTING

```
C      ORTHOGONAL COLLOCATION METHOD FOR SIMULATION OF MIXED
C      IMMOBILIZED CELL REACTOR FOR THE A-B-E PROCESS
C
       DIMENSION DIF1(10),DIF2(10),DIF3(10),ROOT(10),VECT(10),
    *  Y(60)
       COMMON /PAR/ N,A(10,10),B(10,10)
       COMMON /ENG/ SH,DIL,DIF,D,SO
       READ(5,*) N,AL,BET
       READ(5,*) SH,DIL,DIF,D,SO,XO
       WRITE(6,103) N,AL,BET,SH,DIL,DIF,D,SO,XO
103    FORMAT(1X,'N=',I5,1X,'ALFA,BETA=',2F5.2/
    *  1X,'SH,D,DIF,DP,SO,XO=',6E12.5/)
       NT=N+2
       CALL JC CB I(10,N,I,1,AL,BET,DIF1,DIF2,DIF3,ROOT)
       WRITE(6,102) (ROOT(I),I=1,NT)
102    FORMAT(1X,' COLLOCATION POINTS '/1X,8F8.5/)
       DO 1 K=1,2
       DO 1 I=1,NT
       CALL DFCPR(10,N,1,1,I,K,DIF1,DFI2,DIF3,ROOT,VECT)
       DO 2 L=1,NT
       IF(K.EQ.1) A(I,L)=VECT(L)
       IF(K.EQ.1) GO TO 2
       E(I,L)=A(I,L)
       IF(I.EQ.1) GO TO 2
       IF(I.EQ.NT) GO TO 2
       E(I,L)=E(I,L)+VECT(L)/RCOT(I)
2      CONTINUE
1      CONTINUE
       NROV=(N+1)*8+4
       IF(NROV.LE.60)GO TO4
       WRITE(6,100)
100    FORMAT(/' N IS OUT OF ALLOWED DIMENSION RANGE '/)
       STOP
4      Y(2)=1.
       DO 3 I=1,N
       Y(I*8+2)=1.0
3      Y(I*8+1)=XO
       Y(NROV-3)=XO
       Y(NROV-2)=1.
       Y(NROV-1)=XO
       Y(NROV)=1.0
       TF=0.0
       DO 10 I=1,45
       TB=TF
       TF=TB+1.0
       CALL ODEC(NROV,TB,TF,Y,0.01)
       KK=(N+1)*8
       WRITE(6,101) TF,Y(KK+1),Y(KK+2),(Y1(L1),L1=3,8),
    *  (Y(L2),L2=9,KK),Y(KK+3),Y(KK+4),(YN(L3),L3=3,8),
    *  (Y(L4),L4=1,8)
10     CONTINUE
101    FORMAT(/1X,' T= ',F5.2/(1X,8F7.2)
       STOP
```

```
      SUBROUTINE RK4(N,X,Y,H,XE,YE)
      DIMENSION Y(60),YE(60),Z(60),W(60),A(5)
      DATA A /0.5,0.5,1.,1.0,0.5/
      XE=X
      DO 1 K=1,N
1     W(K)=Y(K)
      DO 2 J=1,4
      CALL RHS(N,XE,W,Z)
      XE=A(J)*H+X
      DO 3 K=1,N
      W(K)=Y(K)+A(J)*H*Z(K)
3     YE(K)=YE(K)+A(J+1)*H*Z(K)/3
2     CONTINUE
      RETURN
      END



      SUBROUTINE ODEC(N,XB,XE,Y,EPS)
      DIMENSION Y1(60),Y2(60),Y3(60),Y(60)
      X=XB
      H=(XE-X)/10
C     ADOPTED FROM ALGORITHM 8 C ACM (1960) VOL.3 P.312
C     BY J.VOTRUBA INST.MICROBIOL. PRAGUE CZECHOSLOVAKIA
      IS=0
      IOUT=0
100   IF((X+2.01*H-XE).LT.0.)  GO TO 1
      IOUT=1
      H=(XE-X)/2.
1     CALL RK4(N,X,Y,2,*H,X1,Y1)
101   CALL RK4(N,X,Y,H,X2,Y2)
      CALL RK4(N,X2,Y2,H,X3,Y3)
      ERR=0.0
      DO 2 K=1,N
      C1=AMAX1(1.E-6,ABS(Y3(K)))
      P=ABS(Y1(K)-Y3(K))/Q1
2     ERR=AMAX1(P,ERR)
      IF(ERR.GT.EPS) GO TO 103
      X=X3
      IF(IOUT.EQ.1) GO TO 104
      DO 3 K=1,N
3     Y(K)=Y3(K)
      IF(IS.LT.5) GO TO 4
      H=2.*H
      IS=0
4     IS=IS+1
      GO TO 100
103   H=0.5*H
      IOUT=0
      X1=X2
      DO 5 K=1,N
5     Y1(K)=Y2(K)
      GO TO 101
104   DO 6 K=1,N
6     Y(K)=Y3(K)
      RETURN
      END
```

```
        SUBROUTINE DFOPR(ND,N,NO,N1,I,ID,DIF1,DIF2,DIF3,ROOT,VECT)
        DIMENSION DIF1(ND),DIF2(ND),DIF3(ND),ROOT(ND),VECT(ND)
        NT=NO+N1+N
        IF(ID.EQ.3) GO TO 10
        DO 20 J=1,NT
        IF(J.NE.I) GO TO 21
        IF(ID.NE.1) GO TO 5
        VECT(I)=DIF2(I)/DIF1(I)/2.
        GO TO 20
5       VECT(I)=DIF3(I)/DIF1(I)/3.
        GO TO 20
21      Y=ROOT(I)-ROOT(J)
        VECT(J)=DIF1(I)/DIF1(J)/Y
        IF(ID.EQ.2) VECT(J)=VECT(J)*(DIF2(I)/DIF1(I)-2./Y)
20      CONTINUE
        GO TO 50
10      Y=0.0
        DO 25 J=1,NT
        X=ROOT(J)
        AX=X*(1.-X)
        IF(NO.EQ.0) AX=AX/X/X
        IF(N1.EQ.0) AX=AX/(1.-X)/(1.-X)
        VECT(J)=AX/DIF1(J)**2
25      Y=Y+VECT(J)
        DO 60 J=1,NT
60      VECT(J)=VECT(J)/Y
50      RETURN
        END


        SUBROUTINE JCOBI(ND,N,NO,N1,AL,BE,DIF1,DIF2,DIF3,ROOT)
        DIMENSION DIF1(ND),DIF2(ND),DIF3(ND),ROOT(ND)
        AB=AL+BE
        AD=BE-AL
        AP=BE*AL
        DIF1(1)=(AD/(AB+2.)+1.)/2.
        DIF2(1)=0.0
        IF(N.LT.2) GO TO 15
        DO 10 I=2,N
        Z1=I-1
        Z=AB+2.*Z1
        DIF1(I)=(AB*AD/Z/(Z+2.)+1.)/2
        IF(I.NE.2) GO TO 11
        DIF2(I)=(AB+AP+Z1)/Z/Z/(Z+1.)
        GO TO 10
11      Z=Z*Z
        Y=Z1*(AB+Z1)
        Y=Y*(AP+Y)
        DIF1(I)=Y/Z/(Z-1.)
10      CONTINUE
15      X=0.0
        DO 20 I=1,N
25      XD=0.0
        XN=1.
        XD1=0.0
        XN1=0.0
        DO 30 J=1,N
```

```
        XP=(DIF1(J)-X)*XN-DIF2(J)*XD
        XP1=(DIF1(J)-X)*XN1-DIF2(J)*XD1-XN
        XD=XN
        XD1=XN1
        XN=XP
30      XN1=XP1
        ZC=1.
        Z=XN/XN1
        IF(I.EQ.1) GO TO 21
        DO 22 J=2,I
22      ZC=ZC-Z/(X-ROOT(J-1))
21      Z=Z/ZC
        X=X-Z
        IF(ABS(Z).GT.1.E-5) GO TO 25
        ROOT(I)=X
        X=X+0.0001
20      CONTINUE
        NT=N+NO+N1
        IF(NO.EQ.0) GO TO 35
        DO 31 I=1,N
        J=N+1-I
31      ROOT(J+1)=ROOT(J)
        ROOT(1)=0.0
35      IF(N1.EQ.1) ROOT(NT)=1.
        DO 40 I=1,NT
        X=ROOT(I)
        DIF1(I)=1.
        DIF2(I)=0.0
        DIF3(I)=0.0
        DO 40 J=1,NT
        IF(J.EQ.I) GO TO 40
        Y=X-ROOT(J)
        DIF3(I)=Y*DIF3(I)+3.*DIF2(I)
        DIF2(I)=Y*DIF2(I)+2.*DIF1(I)
        DIF1(I)=Y*DIF1(I)
40      CONTINUE
        RETURN
        END

        SUBROUTINE RATE(R,SS,B,X)
        DIMENSION R(7)
        S=0.0
        IF(SS.GT.0.0)S=SS
        FS=S/(S+2.)
        FI=6./(6.+B)
        R(1)=-(0.005*S+1.25*FS)*X
        R(5)=0.45*0.97/(0.97+B)*FS*X
        R(2)=0.367*FS*X-0.831*R(5)
        R(6)=0.225*FS*FI*X
        R(3)=0.187*FS*X-0.5*R(6)
        R(4)=0.033*FS*X
        RETURN
        END

        FUNCTION AMI(S,B)
        AMI=0.44*S*6./(S+2.)/(6.+B)
        IF(S.LE.0.0)AMI=0.0
```

```
          RETURN
          END


          SUBROUTINE RHS(NN,T,Y,F)
          DIMENSION F(60),Y(60),RT(7)
          COMMON/ENG/ SH,DIL,DIF,D,SO /PAR/ N,A(10,10),B(10,10)
          COMMON/POC/Y1(8),YN(8)
          PSI=14400.*DIF/D**2
          ABET=21600.*SH*DIF/D**2
          DT=A(1,1)*(A(N+2,N+2)+SH/2.)-A(N+2,1)*A(1,N+2)
          DO 1 I=3,8
          S1=0.0
          S2=SH*Y(I)/2.
          DO 2 J=1,N
          S1=S1-A(1,J+1)*Y(I+8*J)
    2     S2=S2-A(N+2,J+1)*Y(I+8*J)
          Y1(I)=(S1*(A(N+2,N+2)+SH/2.)-S2*A(1,N+2))/DT
    1     YN(I)=(S2*A(1,1)-S1*A(N+2,1))/DT
          CALL RATE(RT,Y(3),Y(4),Y(1))
          F(1)=0.034*Y(11+N*8)+(0.56*(Y(2)-1.)-DIL)*Y(1)
          F(2)=(AMI(Y(3),Y(4))-0.56*(Y(2)-1.))*Y(2)
          DO 3 I=1,6
    3     F(I+2)=ABET*(YN(I+2)-Y(I+2))+RT)I)-DIL*Y(I+2)
          F(3)=F(3)+DIL*SO
          DO 10 L=1,N
          K=L*8
          L1=L+1
          F(K+1)=0.56*(Y(K+2)-1.)*Y(K+1)-0.034*Y(K+1)
          F(K+2)=(AMI(Y(K+3),Y(K+4))-0.56*(Y(K+2)-1.))*Y(K+2)
          CALL RATE(RT,Y(K+3),Y(K+4),Y(K+1))
          DO 11 M=1,6
          S1=PSI*(B(L1,1)*Y1(2+M)+B(L1,N+2)*YN(2+M))+RT(M)
          DO 12 I=1,N
          NQ=2+M+I*8
    12    SI=SI+PSI*B(L1,I+1)*Y(NQ)
    11    F(K+M+2)=S1
    10    CONTINUE
          J=N*8+9
          F(J)=0.56*(Y(J+1)-1.)*Y(J)-0.034*Y(J)
          F(J+2)=0.56*(Y(J+3)-1.)*Y(J+2)-0.034*Y(J+2)
          F(J+1)=(AMI(Y1(3),Y1(4))-0.56*(Y(J+1)-1.))*Y(J+1)
          F(J+3)=(AMI(YN(3),YN(4))-0.56*(Y(J+3)-1.))*Y(J+3)
          RETURN
          END
```

TABLE 7.19

COMPUTER PROGRAM LISTING

```
C       ORTHOGONAL COLLOCATION METHOD FOR SIMULATION OF FIXED-BED
C       IMMOBILIZED CELL REACTOR FOR ACETON BUTANOL PROCESS
C
        DIMENSION DIF(1)(10),DIF2(10),DIF3(10),ROOT(10),VECT(10)
   *    Y(120)
        COMMON /PAR/ N,A(10,10),B(10,10)
        COMMON /ENG/ TAU PE,DL,EPS,DIL,SO
        COMMON/POC/ Y1(8),YN(8)
        READ(5,*) N,AL,BET
        READ(5,*) TAU,PE,DL,EPS,DIL,SO,XO
        NT=N+2
        CALL JCOBI(10,N,1,1,I,AL,DIF1,DIF2,DIF3,ROOT)
        WRITE(6,102) (ROOT(I),I=1,NT)
102     FORMAT(1X,' COLLOCATION POINTS '/1X,8F8.5/)
        DO 1 K=1,2
        DO 1 I=1,NT
        CALL DFOPR(10,N,1,1,I,K,DIF1,DIF2,DIF3,ROOT,VECT)
        DO 2 L=1,NT
        IF(K.EQ.1) A(I,L)=VECT(L)
        IF(K.EQ.1) GO TO 2
        E(I,L)=A(I,L)
        E(I,L)=DIL*(VECT(L)*DL/PE-B(I,L)
2       CONTINUE
1       CONTINUE
        NROV=N*13+14
        IF(NROV.LE.120) GO TO4
        WRITE(6,100)
100     FORMAT(/' N IS OUT OF ALLOWED DIMENSION RANGE '/)
        STOP
4       DO 3 I=1,NROV
        Y(I)=0.0
        IF(I.GT.12*N) Y(I)=XO
        IF(I.GT.(13*N+2))Y(I)=0.0
3       CONTINUE
        TF=0.0
        DO 10 I=1,120
        TB=TF
        TF=TB+1.0
        CALL ODEC(NROV,TB,TF,Y,0.01)
        L1=13*N+3
        L2=L1+5
        WRITE(6,101) TF
        WRITE(6,105) Y(12*N+1),(Y(J),J=L1,L2),(Y1(J),J=1,6)
        DO 22 II=1,N
        L1=12*N+II+1
        L2=12*(II-1)+1
        L3=L2+II
22      WRITE(6,105) Y(L1),(Y(J),J=L2,L3)
        L1=NROV-5
        WRITE(6,105) Y(13*N+2),(Y(J),J=L1,NROV),(YN(J),J=1,6)
10      CONTINUE
101     FORMAT(/1X,' TIME= ',F5.2/)
105     FORMAT(1X,F8.3,2X,6F7.2/11X,6F7.2)
        STOP
```

```
        SUBROUTINE RK4(N,K,Y,H,XE,YE)
        DIMENSION Y(120),YE(120),Z(120),W(120),A(5)
        DATA A /0.5,0.5,1.,1.0,0.5/
        XE=X
        DO 1 K=1,N
        YE(K)=Y(K)
1       W(K)=Y(K)
        DO 2 J=1,4
        CALL RHS(N,XE,W,Z)
        XE=A(J)*H+X
        DO 3 K=1,N
        W(K)=Y(K)+A(J)*H*Z(K)
3       YE(K)=YE(K)+A(J+1)*H*Z(K)/3.
2       CONTINUE
        RETURN
        END

        SUBROUTINE JCOBI(ND,N,NO,N1,AL,BE,DIF1,DIF2,DIF3,ROOT)
        DIMENSION DIF1(ND),DIF2(ND),DIF3(ND),ROOT(ND)
        AB=AL+BE
        AD=BE-AL
        AP=BE*AL
        DIF1(1)=(AD/(AB+2.)+1.)/2
        DIF2(1)=0.0
        IF(N.LT.2) GO TO 15
        DO 10 I=2,N
        ZI=I-1
        Z=AB+2.*Z1
        DIF1(I)=(AB*AD/Z/(Z+2.)+1.)/2.
        IF(I.NE.2) GO TO 11
        DIF2(I)=(AB+AP+Z1)/Z/Z/(Z+1.)
        GO TO 10
11      Z=Z*Z
        Y=Z1*(AB+Z1)
        Y=Y*(AP+Y)
        DIF2(I)=Y/Z/(Z-1.)
10      CONTINUE
15      X=0.0
        DO 20 I=1,N
25      XD=0.0
        XN=1.
        XD1=0.0
        XN1=0.0
        DO 30 J=1,N
        XP=(DIF1(J)-X)*XN-DIF2(J)*XD
        XP1=(DIF1(J)-X)*XN1-DIF2(J)*XD1-XN
        XD=XN
        XD1=XN1
        XN=XP
30      XN1=XP1
        ZC=1.
        Z=XN/XN1
        IF(I.EQ.1) GO TO 21
        DO 22 J=2,I
22      ZC=ZC-Z/(X-ROOT(J-1))
21      Z=Z/ZC
```

```
        X=X-Z
        IF(ABS(Z).GT.1.E-5) GO TO 25
        ROOT(I)=X
        X=X+0.0001
20      CONTINUE
        NT=N+NO+N1
        IF(NO.EQ.O) GO TO 35
        DO 31 I=1,N
        J=N+1-I
31      ROOT(J+1)=ROOT(J)
        ROOT(1)=0.0
35      IF(N1.EQ.1) ROOT(NT)=1.
        DO 40 I=1,NT
        X=ROOT(I)
        DIF1(I)=1.
        DIF2(I)=0.0
        DIF3(I)=0.0
        DO 40 J=1,NT
        IF(J.EQ.I) GO TO 40
        Y=X-ROOT(J)
        DIF3(I)=Y*DIF3(I)+3.*DIF2(I)
        DIF2(I)=Y*DIF2(I)+2.*DIF1(I)
        DIF1(I)=Y*DIF1(I)
40      CONTINUE
        RETURN
        END

        SUBROUTINE ODEC(N,XB,XE,Y,EPS)
        DIMENSION YI(120),Y2(120),Y3(120),Y(120)
        X=XB
        H=(XE-X)/10.
C       ADOPTED FROM ALGORITHM 8 C ACM (1960) VOL.3 P.312
C       BY J. VOTRUBA INST.MICROBIOL. PRAGUE CZECHOSLOVAKIA
        IS=O
        IOUT=O
100     IF((X+2.01*H-XE).LT.0) GO TO 1
        IOUT=1
        H=(XE-X)/2.
1       CALL RK4(N,X,Y,2.*H,X1,Y1)
101     CALL RK4(N,X,Y,H,X2,Y2)
        CALL RK4(N,X2,Y2,H,X3,Y3)
        ERR=0.0
        DO 2 K=1,N
        Q1=AMAX1(1.E-6,ABS(Y3(K)))
        P=ABS(Y1(K)-Y3(K))/Q1
2       ERR=AMAX1(P,ERR)
        IF(ERR.GT.EPS) GO TO 103
        X=X3
        IF(IOUT.EQ.1) GO TO 104
        DO 3 K=1,N
3       Y(K)=Y3(K)
        IF(IS.LT.5) GO TO 4
        H=2.*H
        IS=O
4       IS=IS+1
        GO TO 100
103     H=0.5*H
        IOUT=O
```

```
      X1=X2
      DO 5 K=1,N
5     Y1(K)=Y2(K)
      GO TO 101
104   DO 6 K=1,N
6     Y(K)=Y3(K)
      RETURN
      END

      SUBROUTINE DFOPR(ND,D,NO.N1,I,ID,DIF1,DIF2,DIF3,ROOT,VECT)
      DIMENSION DIF1(ND),DIF2(ND),DIF3(ND),ROOT(ND),VECT(ND)
      NT=NO+N1+N
      IF(ID.EQ.3) GO TO 10
      DO 20 J=1,NT
      IF(J.NE.I) GO TO 21
      IF(ID.NE.1) GO TO 5
      VECT(I)=DIF2(I)/DIF1(I)/2.
      GO TO 20
5     VECT(I)=DIF3(I)/DIF1(I)/3.
      GO TO 20
21    Y=ROOT(I)-ROOT(J)
      VECT(J)=DIF1(I)/DIF1(J)/Y
      IF(ID.EQ.2) VECT(J)=VECT(J)*(DIF2(I)/DIF1(I)-2./Y)
20    CONTINUE
      GO TO 50
10    Y=0.0
      DO 25 J=1,NT
      X=ROOT(J)
      AX=X*(1.-X)
      IF(NO.EQ.0) AX=AX/X/X
      IF(N1.EQ.0) AX=AX/(1.-X)/(1.-X)
      VECT(J)=AX/DIF(J)**2
25    Y=Y+VECT(J)
      DO 60 J=1,NT
60    VECT(J)=VECT(J)/Y
50    RETURN
      END

      SUBROUTINE RHS(NN,T,Y,F)
      DIMENSION F(120),Y(120),RT(7)
      COMMON/PAR/ N,A(10,10),B(10,10)
      COMMON/ENG/ TAU,PE,DL,EPS,DIL,SO
      COMMON/POC/Y1(8),YN(8)
      DT=(A(1,1)-PE/DL)*A(N+2,N+2)-A(N+2.1)*A(1,N+2)
      TAU=0.1
      PE=2.
      DL=1/20.
      EPS=0.5
      DIL=0.15
      SO=40.
      DO 1 I=1,6
      S1=0.0
      S2=0.0
      IF(1.EQ.1) S2=PE/DL*SO
      DO 2 J=1,N
      JJ=J*12+I-6
      S2=S2+A(1,J+1)*Y(JJ)
2     S1=S1-A(N+2,J+1)*Y(JJ)
```

```
       YN(I)=(S1*(A(1,1)-PE/DL)+S2*A(N+2,1))/DT
1      Y1(I)=-(S1*A(1,N+2)+S2*A(N+2,N+2))/DT
       F(12*N+1)=-0.0014*Y(12*N+1)*Y1(2)
       F(13*N+2)=-0.0014*Y(13*N+2)*YN(2)
       DO 3 I=1,N
       J1=I2*N+1+I
       J2=(I-1)*12+2
3      F(J1)=-0.0014*Y(J2)*Y(J1)
       DO 10 I=1,N
       J1=(I-1)*12
       J2=12*N+1+I
       CALL RATE(RT,Y(J1+1),Y(J1+2),Y(J2),Y(J1+6),Y(J1+5))
       DO 11 J=1,6
       J3=J+J1
       J4=J3+6
       F(J3)=RT(J)-(Y(J3)-Y(J4))/TAU
       F(J4)=(1.-EPS)/EPS*(Y(J3)-Y(J4))/TAU+B(I+1,1)*Y1(J)
*      +B(I+1,N+2)*YN(J)
       DO 12 K=1,N
       K1=(K-1)*12+J+6
12     F(J4)=F(J4)+B(I+1,K+1)*Y(K1)
11     CONTINUE
10     CONTINUE
       CALL RATE(RT,Y1(1),Y1(2),Y(N*12+1),Y1(6),Y1(5))
       DO 20 I=1,6
       J1=N*13+2+I
20     F(J1)=RT(I)-(Y(J1)-Y1(I))/TAU
       CALL RATE(RT,YN(1),YN(2),Y(N*13+2),YN(6),YN(5))
       DO 21 I=1,6
       J1=N*13+8+I
21     F(JI)=RT(I)-(Y(J1)-YN(I))/TAU
       RETURN
       END

       SUBROUTINE RATE(R,SS,B,X,AA,BA)
       DIMENSION R(7)
       S=0.0
       IF(SS.GT.0.0)S=SS
       FS=S/(S+2.)
       FI=0.833/(0.833+B)
       R(1)=-(0.024*S+0.6*FS)*X
       R(5)=(0.0135*S*FI-0.1*BA/(BA+0.5))*X
       R(2)=0.0107*S*X-0.831*R(5)
       R(6)=(0.22*FS*FI-0.01*AA/(AA+0.5))*X
       R(3)=0.132*FS*X-0.5*R(6)
       R(4)=0.026*FS*X
       RETURN
       END
```

INDEX

Printed and bound by CPI Group (UK) Ltd, Croydon, CR0 4YY

03/10/2024

01040330-0007